RIVER ETERNAL

RIVER ETERNAL

LANCE KINSETH

VIKING

VIKING
Published by the Penguin Group
Viking Penguin Inc., 40 West 23rd Street,
New York, New York 10010, U.S.A.
Penguin books Ltd, 27 Wrights Lane,
London W8 5TZ, England
Penguin Books Australia Ltd, Ringwood,
Victoria, Australia
Penguin Books Canada Ltd, 2801 John Street,
Markham, Ontario, Canada L3R 1B4
Penguin Books (N.Z.) Ltd, 182–190 Wairau Road,
Auckland 10, New Zealand

Penguin Books Ltd, Registered Offices:
Harmondsworth, Middlesex, England

First published in 1989 by Viking Penguin Inc.
Published simultaneously in Canada

1   3   5   7   9   10   8   6   4   2

LIBRARY OF CONGRESS CATALOGING IN PUBLICATION DATA
Kinseth, Lance.
River eternal.
1. Life.   2. Philosophy of nature.   I. Title.
BD431.K49   1989   113   88–40288
ISBN 0 670 82441 0

Printed in the United States of America
Set in Sabon
Designed by Sarah Vure

for Lynne

ACKNOWLEDGMENTS

There are four graces that honor my life and that have presence in *River Eternal*: the *joy* of life of my dearest companion, my wife, Lynne; the energetic *wonder* of discovery of our daughters, Kristin and Kathryn; my mother's *affirmation*, which, early on, moved me forward against the odds; and my father's deep *affection* for the natural landscape, which has given focus to my journey.

Of immediate significance to the publication of *River Eternal*, my gratitude extends, first, to writer and editor Robley Wilson, Jr., for his initial support and direction, and for publication of an earlier version of the first chapter, "Homing," in *The North American Review*. I am also indebted to Barry Lopez for his early and continued encouragement. Finally, my deep appreciation extends to my editor at Viking, Dan Frank, for his remarkably sensitive provocations, which have resulted in a more lucid account.

Each word of *River Eternal* is far more than mine. Each has been scrawled by a body without end. I have been nurtured by a strong current of literature that continues to illustrate in many local landscapes, in mountains and seacoasts and forests and deserts, just what miracle we inhabit and express. And, of course, this account would have been impossible without the river and its expressions, crow and heron and willow and turtle and the fine grains of sand. In all places there is this profound opportunity to advance beyond the narrow latitude and longitude of our separation and dominion.

In writing this account of a prairie river, I have taken only the tea-leaf patterns at the bottom of the bowl. The taste is still there in this rivervalley, and in your local place as well, awaiting you.

C O N T E N T S

ix

RIVER ETERNAL

An Invitation

I offer you a story about yourself. It wears the appearance of a praise of River, of Earth, of Star. In your stolen moments grasp the tail of this praise and be lifted out of the strong current of your everyday. Fly inside the solitude of still pools and the calming staccato of rapids. I will give you the dazzling, never-subsumed molten colors of common and ashen days.

I am going to fly so deep inside the dance of a river that your everyday may be changed forever, becoming perhaps more subtly colored or perhaps being more noticeably lost to a flood of sensations. To the degree that you are willing to risk, recover then a sensibility that always endures inside you but that can be lost or stray from or be stolen outright from your everyday. Discover your life inside these waters. Here is a way to live an enchanted life in the midst of the everyday, where every moment can open on Always.

Your everyday can be a daydream, a stage set, a sleepwalk. Through the window glass, a bird sings. You ask how this view could be more real, and if it could be, why more than this? A connection has been made between a bird and yourself, and still, there is a disconcerting sense that you have contrived a wall more than you have opened a doorway. You stand *within* a landscape, and yet you continue to look *at* it, as if you were separate from it and the landscape were, at best, another nation.

Such an everyday view is not as wrong as it is incomplete. The exactness of a bird, the detailing that makes you ask how there could be more than this view, is only a thin veneer. There is so much more. You do not have to take my word for it. You can trust your own intuition. Here is Aldo Leopold's "numeron," that eternal essence underlying the measurable "phenomenon." And here is your original face, yes, that one that Zen patriarch Hui-neng questioned you about, that one you possessed before your father and mother were born, and that one that will sustain for the long run of geological time.

The bird outside your window is so much more than another complete nation. You are drawn by an implication of familiarity, of connection. Each appearance is an expression of something larger, as if that appearance were a visible whorl in a stream. Every appearance is less an object and more an invitation, a point of entry. This changing view is an opportunity for cultural advancement, for enlightened movement forward, for an expansion of a too-narrow, sleepy view of self and place.

This changing view is the same world of your everyday but with a hairsbreadth difference. And I am going to let the cat out of the bag right now, before the story really begins. To go inside this changing view, simply shift your awareness to the *activity* of appearances, to the flight that carves the feather. If you could stay with the activity of just one event forever, it would fly in an unbroken, inseparable current back inside everything everywhere, including yourself. In a very real way, a bird, and even yourself, expresses a star. You can find this bird and yourself to be an array of atoms born in ancestral stars, and deep inside the evolution of one star in the here and now. Each event is a landscape that extends into infinities of largeness and smallness. And by this view, you recover this miraculous, deeper working body of yourself.

There is, in and surrounding you, a skin that wears like light. It can stretch finer than a frog's bellowing throat. It is as fluid as every change in the weather. Cobwebbed and silken, it extends inseparably forward, flying between the walls and through the willows and under the hill. If you want a less veiled glance at it, your shadow is a place to begin: It is yours and also outside you. But if you are looking for its wide-open dance in the landscape, there is, of course, the knockout magic of all rivers.

If you are looking for a way inside this changing view, for a new key of sorts, there is a way opened by the direct experience of a landscape. The remarkable appeal of all lands is their profound capacity to complete you. All landscapes afford the ultimate intimacy of being not simply in touch but *inside*. And when you go inside that which can cut a city in half—a river—you are inside a landscape that is more than enough to overpower your limitations. By a river's fluidity, you can be lifted out of gravity into the midst of a wild, splitting, and fusing terrain. And the thoughts that rise out of the meld of your contemplation and the river's purl can be like sheet lightning—whole, electrifying, purifying—coloring deeply that part of your thinking that is still a wild young shout.

By a river's continual change, by its dependable flow forward, I am steadied. The river offers something more compelling than personal recovery. It offers something beyond my private moment to that deep stream of our species to come. When I write to you about the river, I am not writing another nature travelogue to an "out-there." This is environmentalism, an attempt to override the treason of the belief that culture and nature are separate. When I write to you about the river, I am writing to you about the long part of ourselves. The great joy of this changing view is that we can never be lost or isolated. No matter where I stand or how much the local place changes, and no matter how far I wander, there is only constancy in new clothing.

My journey into the events of the river likely began as a nature story, a travelogue to other nations. However, by my continued returns, a sense of separation could not be upheld. Place—landscape and environment—began to be experienced as the long part of myself. The story of a river, and of all landscapes, becomes in a very real way the autobiography of each of us. The landscape becomes not only the deep prologue to culture but also our immediate story and the deep economy of our future as well. Every one of our breaths and every one of our words are small glimmers of its streaming dance.

Inside a river's forward-reaching, I am swept into a view that is both deeply rooted and endlessly refreshing. A river is a long view—a penetrating line that swims back to its living origin and sweeps ahead into its future, remaining in touch with it all. There is a rivering of sorts in everything, in the belly of a valley—in the physical river, of course—and in the microecology of the veining of every leaf and across the back of my hands and inside of my looking eyes, in the atmosphere and in the ocean—yes, in the middle of all that water—and in the astroecology of galaxies.

In every local landscape, and especially in everyone's remembered river, I have come to believe that we can stumble upon our pilgrim ship and its glory road, all wrapped into one. Here is the elixir for which we have been willing to search worlds to discover. We have believed cultural evolution to be a process of escaping the violence and chaos of a tooth-and-claw nature to invent peace. Our aggression toward ourselves and our degradation of the landscape have been perceived to be the vestiges of our animality. I believe that we have deluded ourselves. In all the overlooked, discarded local places—along a river's edge and coastline, in a small copse of wood and in a meadow—we stand to find the structure of a long-standing peace. This peace is unnoticed in the indigenous landscape not because it is rare but because it is too successful. It is like the less dramatic, enduring, good news story.

Our way out of a growing global desert, out of our warring with ourselves and with the landscape may be simpler than we could have ever imagined. If we will bring to these landscapes our sensibility and not just our measures, we can come to resemble the harmony that we discover there. As Alfred North Whitehead writes in *The Adventure of Ideas,* "The experience of Peace is largely beyond the control of purpose. It comes as a gift . . ." Perhaps when we can seriously honor the overlooked events of the landscape, we will have developed both the wisdom and the will to receive this gift. In all landscapes there is that powerful commonness that holds the world together. It is river eternal, the streaming beyond appearances, the inseparability that makes one thing also the other, that is nothing but cooperation and fittedness.

River Eternal is a selection of my direct experiences of the activities of a prairie river. It begins with a prologue, "Before Taking One Step," to set the directive for a homeward journey. The first chapter, "Homing," introduces the fluid metaphor of "rivering" and goes inside the rich imagery of a common day and night. "Streaming" challenges the mask of appearances to reveal a more fundamental form underlying appearances. "Passage" approaches the fluid openness of time and space. "Going Inside" explores my experience of inseparability from the landscape that congeals into the metaphor of river eternal. An epilogue, "The Voice in the Willow Quarter," offers one last scenario to realize the longer body of self—a fictional voice of the landscape. The epilogue is a point of departure, a way to leave the river and still carry it within as a voice that can continue to speak to you in your everyday.

Lean into a favored tree or into the sweet grass of the hill.

Relax and be drawn forward into these pages as if rafting effortlessly downstream. Drift and be open to the possibilities evoked by each turning page. Let this story be a provocation, a trigger, more than an answer. Be outward bound to a river's dance and seed an inner fire. Let it become a soft rain of waking light. Suddenly wake inside a landscape of delicacy and power.

LANCE KINSETH
goat-hill prairie remnant
S.½., N.½, Sec. 22, T.88N, R.28W, 42.2N—94.07W

Before Taking One Step

To complete ourselves, we have been willing to set out for the far corners of the world. In rarity, we hope to be shocked into self-realization. And yet, for all the potential of the exotic to challenge us and extend our conceptual range, it is more likely to bewitch us and distract us, keeping us always looking someplace else. In seeing the world as distant from ourselves, we keep it from ourselves no matter how far we wander. And in not seeing the local place where we stand, what can we hope to discover someplace else? We may make distant journeys in hopes of looking back. And yet, to complete ourselves, we do not travel as much to distant worlds as to that which we distance from ourselves. More than getting to someplace else, the essence of travel involves bringing the world inside our identity. And this journey begins before taking one step.

Any local place is a far-off world, as difficult to reach as a star, distanced not by physical space but by barriers of habit. Throughout cultural evolution the lure of the local place has been diminished by our increasing capacity to wander physically farther and faster as well as to extend our sensory awareness beyond the reaches of our physical mobility. There always seems to be too much else to do before such a journey to the local place can be seriously pursued. When we turn to the local place, it is either to pick the bone clean, extracting some remnant material resource, or to take brief respite from our "real" work. In our most intimate relationship with a local place, we celebrate loss, as John Cheever writes in *Oh What a Paradise It Seems,* "searching for the memory of some place . . . as if it were some country which we had been forced to leave." We feel closer to nature, and yet we are grateful for our exile, like immigrants having found a more bountiful land in culture and attempting only to keep a memory alive.

Attention to the local place seems a step backward, in many ways an avoidance of the world and a retreat from a search for solutions. And yet, by extending our scientific awareness to light-years distances, we have paradoxically discovered the depth of where we stand. In a time when nature has appeared to be so far from civilization, we find ourselves deeply immersed in the cosmos. And importantly, we begin to encounter the presence of this whole nature in every local place. We begin to revision the local place as no longer simply prologue or background to our journey but the journey itself. We begin to sense that with every step we take from our local place, we stand to leave the world behind. Thoreau's seemingly impossible paradox—to travel a good deal in one place—becomes not simply a curious intellectual journey but perhaps *the* strategy for survival.

If we were given the gift of traveling anywhere we might choose, would we have the wisdom to stand in one place?

Homing

It could be the beginning of one day in any season of the year. I have come seventy miles north of one of the wildest events in nature—my capital city's wake-up rush. I have put the car to sleep high on the hill and have sauntered for nearly a mile through an upland gallery of hickory and oak. My pace has gradually become that of the terrain's—a down-streaming pattern of crosscuts and stops and go-arounds. It is just the start of the slow roll of one morning. Suddenly, the forest opens onto a steep prairie ridge overlooking a river valley. The dominant view is of rivers of sky.

I am standing on a steep goat-hill vista. It is a razor's edge lost among soft-shouldered, over-old forested hills. And it is just one small notch in the brim of a 500-mile-long stream. And that stream is, in turn, just a small crack in a vast 250-million-acre table of midcontinent. Inside this small notch I have encountered a deep economy and a grace that has drawn me again and again like steel drinking in the force of a magnet. I have spent a lifetime of seasons in this notch, and it only more deeply intrigues me. And so I have come again—an anchorite seeking to be steadied inside the permanence that this landscape always affords me.

I stand at midpoint, midcontinent, in the heartland. It is as rich as a Mark Twain landscape. There is a Wildcat Cave and a Boneyard Hollow. There are sandstone and limestone outcroppings and river islands and elbows, and odorous beds of copperas and paleo-Indian mounds. There are cliff swallows and turkey vultures and flying squirrels and delicate ferns and liverworts, and a thousand more knockout curiosities that assemble into this complex river valley. This valley is a wish come true, without having dropped a penny in the fountain.

The first sound that I hear on this goat-hill is emptiness, or perhaps more of an emptying. There is the sound of no-car, no-plane, and no-word. And my first view is the tops of everything, and they shrink beneath the heavens. It is a cloud's perspective, and I cannot begin to subsume it. But I have been here before and I know what's coming.

Gradually the structure of stillness opens. There is the jump of the wren's jam-packed rattle and the woodpecker sounding the Devil's Tattoo and, far-off and high on the hill, my special ally, the bark of one crow alerting the valley and acknowledging my arrival. The sky relaxes its grip on my senses. I begin to become aware of the way that the hills roll and fall and cast about and then ascend. The colors of the forest open to a chorale. Upstream, a plodding settlement, a coal town, waltzes with a sleepy energy.

No matter how far I descend into the details, I can find this music that looks like wonder and deep economy. On this goat-hill vista, under my feet, big and little bluestem and wild rye and switchgrass, poverty oat grass and side oats grama form assemblages, interspersed with the blooms of prairie clover and purple coneflowers, red-orange butterfly milkweed, leadplant and goat's rue. Appearing to be only a vestige of the past, this prairie is perhaps a model for the city of the future. Here is an event that long ago resolved our current dilemma of living long-term as residents more than migrants. Without thinking about it, this gallery of plants demonstrates the long-run value of stability over disturbance, stronger root systems over high reproduction, and cooperation over dominion. Enduring a worst-case scenario of fire and freezing wind and droughty soil and full sun, this landscape of wisdom banks a sweet profit—a stored capital of soil—that has underpinned civilization.

Beneath and above this goat-hill, there are a thousand other success stories that become listening points for wisdom. Beneath this vista there are nettles and willows and butternuts and maples with exciting tales of how and why they came to live in this place. There are sandstone cliffs, a waterfall, and a cobblestone riffle on the goat-hill side of the river. There is the fluid sweep of the heron and the crane and the fast, rolling curves of the cliff swallows, all flying low over the matt of the river. There are diamonds in the water and the fine-stitching flight of dragonflies and black damselflies in the rushes of the river's brim. There is the shrill of killdeers and the buzz of both wetland blackbirds and tall-grass katydids. Far overhead, one star, meager by cosmic standards, blasts this chamber by day. And by night the valley is revealed to be woven to a thousand other suns.

What draws me to this valley is at the center of its body. It is the wide-open vena cava of the eternal, the river. I descend this steep hill and push my hand inside a vital energy going downstream to be taken to the heart—the sea—and come around again. For all its sensitivity, my hand cannot detect its fine, measured pulse, but there is something that feels like a living miracle should feel. The sensation is one of reaching inside the miles-long roll of a curve. My touch is a long sip of a tea, eloquent enough to satisfy the most astute taste. In this tea-brown river I have a solid name for the grand process of nature, inside of which I find myself as a current melding continuously with other currents of earth and air and fire and water. It is *rivering*.

I glance back up at the rock-firm goat-hill and sense its downwasting, melting like butter in geological time. Eroding far faster than these hills, the earthstream has been easy to miss. And yet this goat-hill is the visible tip of an iceberg, constituted of sandstone and base rock and continent, floating on a river of molten earth core. And the visible rock faces are the thin casts of change, of coal swamps in sandstone and inland seas in limestone, both etched and partially erased by glacial ice. Atoms shower off these rock faces, raining into wind and water to be finally drawn as sediment down the river. On the river's brim I am charged between these poles of stone-slow change and the river's quick rush. I am spun around and lifted out of my gravity into the fluidity, the rivering, of even the rock-firm ground upon which I stand. I am a stream overlaying a stream, and under rivers of sky that are shimmers lost in a cosmic stream.

Wherever I find myself, there are rivers. There are vast continental rivers, of course, and there are regional rivers, rivulets, creeks, and canals. There are lines streaming across boulders and down inside hairline fissures. A river is streaming through every part of the day and night, through a summergreen forest and then through a city, and right through the middle of the ocean, through all that water that is more than water. And right now, wherever I sit, a river percolates through the strata beneath me. There is a rivering—a weaving and streaming of sorts in every appearance. There is a continual coming to and leaving of form, right next to me and even inside me.

Rivering is so much more than water. Forever going down the drain, a river affords a view of the world from a different angle. The undersides of trees and bridges are the obvious revelations in its mirrored surface—the everyday simmering tea that any river serves us effortlessly. Yet a river runs so much deeper than this. It is one elegant tea of gives and takes, and it is an event linked to everything everywhere. Even the river that is rolling through a city gets taken inside the events that occur in that local place, no matter how hard-edged the landscape may suddenly become. And it is especially in this mixing of both imagery and real activity that a river approaches the essence of the world. Rivering turns the most hard-edged landscape into a fluid event. Its surface is a crack between worlds, between the grounded appearances and the ethereal activity. And no matter how narrow this streaming gets, it remains broad enough to tie reason and intuition together. Rivering can open a watershed way of perception, especially opening fluidity of form—the larger nature of process underlying the visible nature of appearances.

River is joining, *riveting,* merging parts so solidly together. Estuary is joined to marsh to creek and this creek to sky and river again and again. At the same time, a river *rives,* splitting one valleyside from the other as well as dissolving sediment. This great contrast assures that just when I begin to sense a river as one thing, there is always a sense of this process of rivering containing more than I had suspected. My foot may tell me what my reason won't. My chosen path might be a mirage. There can be a deep hole in the next step. The ice collapses. My body stands to get lost, to get woven in some unintended way to the landscape.

In just one obscure waterway there are so many perceptual drinks. The river below this goat-hill vista may be one very obscure waterbody in the cosmos and yet it contains as many aspects as a cosmos. Strong-Back, Strong-Arm, Withering Finger, Wild Glint in the Eye, Dancing Light: These are all aspects of a river. Half-circling, elbowing in one place, then straight-running or reaching in another, this too is a river. It has no single name or single dance. It is the China-Breaker and the Naked Throat. In any one obscure waterway, and in any square inch of its volume, there are particles of all seas and the well-sung bodies of all rivers. Here is the holy cow's mouth—Gaumukh, the Ganges' source—the waltzing Danube, the Sumida, the Yukon, the Ob, the Yenisei, the Congo, the Nile, the great Amazon, the Paraná, the Mackenzie, the Leam, and the Godavari. There are so many drinks in this one drink—murky, muscular at times, even torrential and bloody, or suddenly serene and deeply crystalline.

When I try to imagine a world without rivers, I am left standing on a harsh, dusty landscape. A river is one place that is really doing something without trying to do anything. It is common and not eccentric. I learn something from the common, from all its overlooked aspects. For all its commonness, a river is capable of startling me. This event can be destructive as well as nurturing, passive and silent one time, then raucous and outraged at another. Constantly cutting a new figure, I could spend a lifetime in the bend of one obscure river, sensing the daily changes in the configurations of stones in one obscure point. Toiling, a river is farmer and a soldier. Sweet, it is a rubbing hand at the belly. Snaking, it works between the hips. Man-woman, unmanly, unfeminine, emasculated, never human; something so familiar, something so different; open, a tongueless mouth, unheard from, invisible, alone—this is a river, one little thread that is a thread woven to the world and at the same time the edge of a wave.

With rivering going on in all events, it is not essential to go to the physical river to find it. Looking into a leaf, I see a river of veins. I look at my hand, and rivers snake across the back of it. And yet I am drawn to the physical river. For me, it is William Blake's "energy as eternal delight" expressed with clarity and splendor. It looks to me like an open nerve of energy, unencumbered by form, changing, fluid, and in motion. I am drawn less to the clarity it affords, and more by the way it relaxes and steadies me. I can choose still waters or smoothly rolling current or the speeded-up riffles and purge of a flood. I respond to a river as if I were going inside the comfort of home. And I am near the heart of all events—the eternal energy—which washes away any traces of ownership that we put to it and is preeminently inventive.

Around one bend, arums—arrowheads—root at the edge of a water canyon, a deep bowl in the river. The pool is a plump belly, trailing a long tunic well downstream. There is a deep center that I cannot penetrate with my eyes. All this basin, the center and its rooted brim, is like a fine celadon bowl fit for a tea ceremony. Its liquid is more than water, so much more than concepts such as soup or a boiled-down stew which do not match the river's metaphoric power. I sense that it has both the appearance of a rich tea and the promising power of an elixir.

A river's tea color, sometimes milky with sediment, indicates the promise of more complexity than water. However, its base element, water, is not that easily dismissed. The greater body of the earth's surface is an electronic silicon ash, born largely in ancestral volcanoes, and to a small extent now in that which appears to be its opposite, fire. Strike a match and water bursts out of thin air. And each triatomic molecule is blessedly off-balance electrically, with its 104.5-degree angle between its two hydrogen atoms, giving water, and this tea, its "stickiness," which bonds other compounds inside water and promotes complexity. When I sit down by this teabowl, it is before a tea that is wild, chaotic, and wide open.

In this teabowl I have a drink of far more than water to outmatch my thirsting mind set down beside it. Almost every flavor of this drink has been taken apart, magnified, dissolved, assayed, and burnt. And still this drink is virtually unknown. Bottled, its essence strays. And so I am drawn to the river and its tea, which regenerates a strayed sensibility.

The river is a tea alive and vital—a Van Gogh plasma. Even ice-cold, it perks and simmers and charges. Vital like protoplasm, it is elemental and yet a masterwork. Not a Sunday watercolor, a river is a fine art, a Cézanne form within form or a Rothko color field. Turn the bowl and wonder about the terrine itself and then perceptually drink to life: on top—bubbling oxygen and iron oxide form a blue-green film; underneath, a drink of mouths and watery eyes, a drink of children and baby carriers and old sentient beings, an elixir of birthings. Drink of the unliving: a tea of leachings, disassembling and reassembling, and blending. This is a charged tea, a drink of every possibility—thick and thin, hot and icy cold. Biological, unbuilt, and recently contrived with chlorinated hydrocarbons, mercury and arsenic and selenium compounds, and carbamate insecticides—in all, traces of as many as 900 basic chemicals, for better or worse, like the world of today, a tea of everything.

Tea is spilling out of the bowl, going downstream, and passing into another deep vessel. Around one upstream bend comes more fresh tea to fill this emptying bowl to its brim. With this sojourn continuing for centuries, it is amazing that this stream averages only two feet in depth. Not pressing down, this tea streams forward. There are currents in this complicated tea that slow and press down deeply, seeping through a foot of sand to claybed, and press even deeper through bedrock, molecule by molecule. But the great body of currents runs fast. At a turn they lose control and collide and wear the riverbed sideways. As a result, old rivers writhe like slow uncurling smoke. There are some thin lines of current that run so fast that they fly out of the bowl and seep into the rivers of the sky.

The diversity of this tea's composition and its rigid fluidity satisfy my thirst and open a sensibility that sustains into my everyday. Now and again I return to drink of the river and to give a day back to it. Even then, I receive more than I give. A day of my time spent on a river becomes a small season of delight.

Everywhere a day is the spinning of the whole earth. And any one of its events, from the birthing of a mouse litter to the fall of a venerable tree, will swallow all my capacity for understanding. Each day is an opportunity to wade inside a still-molten bullion in a brittle-cold cosmos, and to have for an answer nothing but wide-open possibility. The quiet restraint of so much force is the calming ingredient in the swirl of energy that we name *day*. It is the leading edge of creation, of John Muir's "not yet half made" world. And the grand scope of a day's self-restraint, of a power that dwarfs our pandemics of genocide and degradation, is a model for us to encounter and replicate.

Like the mythical phoenix, a day rises to life from its own ashes. It is the living landscape, and its lesson is the great one: there is no end. It is the immortality that meets the demands of our modernity—that can be touched and tasted. And so I come to spend the full season of one day on my river: just before first light, dawn, morning, and noon, sunset and dusk, and deep night. Well spent, one day is the privilege of a lifetime. And there is one day after another to spend.

In the very earliest moment of dawn, in the most minuscule part of that moment, the overwhelming ambience of the streamvalley is one of smoke. Atomized, vaporous, the valley appears near creation, before any condensation to form. And again the wisdom of the world speaks to me in a voice that I can grasp to wake me to the creation of the moment. The valley hangs for a moment on the brink of darkness. And it seems that only a moist scent keeps it from nothingness. But in another moment this valley will burst into the optical nerve endings and I will fall far behind in my best efforts to keep pace with the messages that are storming inside me. For at the very moment of dawn the valley will congeal fast as a crystal on a string and rise into form in the cool morning sky.

Just before dawn the sky and the river have appeared to be fused. The river overperks, spilling out over the rim of the valley floor, and rises into the alluvial fields. Everything is fog and, moistened by it, takes on the appearance of new life. Fog is the river going up into the rafters. And when it occurs, dust in my throat no longer feels like dust in the throat. My breath is revealed to be vaporous, forming small currents and eddies before my eyes. It seems as though a fish might pull up next to me as if I were a boulder in midstream.

Even though my eyes are dulled by it, the fog opens up the concept of rivering. The fog is comforting and relaxing, steaming like a slow-brewing kettle. It feels and smells like a river might to a fish, and it is supportive in the way that it feels skintight and riveted to me. A tree vanishes, then reappears. In my perception this tree is taken apart for a second and then reassembled. Molecules come to mind more than objects, and there is activity in the space between objects that my eyes had sensed as emptiness. The world is seen whole and inseparable—a field of currents, with whorls becoming visible as a storm of molecules or color or condensed energy and then washing back into their larger body.

Nuclear fusion energy missiles over the horizon. My eyes are struck by needle-thin strips of light. This is dawn. A bomb has gone off nearly one hundred million miles away, and a bit of it is ramming into this streamvalley. I always forget. I never subsume it. I am always expecting the first light to be a dim light, dusty and cobwebbed, like a bulb burning off in a cellar. But this first light is a blasting light. It seems to crack the dark where it enters, radiating out like splintered glass around a bullet hole. This is a light at which I cannot even look.

Dawn is a warming fire, thawing out the damp, the chill, the creaking joints. It seems to shatter into a thousand pieces on this river, with particles skipping across the top, then melting in. True, the sun comes harsh as a hammer through a door, but it gets welcomed inside every day. What a thing it is to wake up every morning to a fusion bomb that the world waits for and depends upon!

The sun is going full blast right out of bed. It thrusts a finger through the hair and it will come through the smallest crack in the door. It can be personified, given a finger to thrust with or put to bed at dusk, but the sun is indifferent, so immense and yet able to squeeze magically inside a hairline fissure and become plants and animals. It is too much for my eyes and it is strong enough to gag a forest and dry a river. I know this for a fact. I am always finding the bones of an old river in stone and infant river in mud, drying and cracking in late summer.

This fission bomb is so precious. Dawn makes the river into a pot of gold. Gray bark turns gold for an instant. The sun is giving me a glimpse of the preciousness of the valley in color terms (e.g., gold) that I can understand. Everything is turning to gold at dawn, uncorroding, unblemished, and bold. The sky has become a dome of gold. There is gilt everywhere: gold on water and meadow, aurum rays straight as arrows leaning up against trees in the upland forest. Gold has cut through the shroud of the night. For a moment, the streamvalley at dawn is heavy and dense, glittering and weighty. Here is alchemy, a goldstone for contemplating that easily outweighs the metal itself.

The sun dissipates the fog but in a remarkably silent way. The way in which the visual world suddenly snaps together makes me feel as though I ought to wake up to a sharp crack. Suddenly the grasses are fixed where they were yesterday and the "objects" seem to put a vise and clamp on the ambiguity of night. If there is a sound to dawn, it is one bird. And then this sets off another. Their eyes peer at the sun as it cracks the brim of the horizon and beams into the forest at a stiff angle. A path, narrow yet well-worn, collects small pools of light. One side is spattered with light; the other seems to rot in darkness. The birds make a music to fit this kind of world, chirping in small bits and pieces, as if the notes were clipped with fine scissors. It could have been a blast, but it is nowhere as harsh as a revelry.

This dance of light and shadow gradually inches into the sky. And this waterless star sets off the plumbing along this path with each ascending step. Here comes a river of water *up* through the capillaries of the plant at a rate of an inch per second. Millions of plants respond, each according to its relationship to the sun. Each plant is braiding itself to the sky and to the sun. And paradoxically, like the veil of fog, the ascent of the sun begins to reveal the inseparability of one event from another as I begin to observe the activity of the new day. Walking for years through the local place, I have seen the object. But now, as I stay with the object rather than physically move about, I begin to see its activity, and my journey is a deeper one.

Throughout the ascent of morning, this river is a chameleon—rose, then amber, then rapidly catching up with the sky that is high-white. Focus closer and closer upon the river: Water striders skate upstream against the current, treading on cloud, bird, the images of overhanging branch and stem, and butterflies and leaves and small flies. Every square inch of the surface is changing. In every minute drop of this river, light is flying and tumbling like a swallow, going over foam, bubble half-domes, shell and seed particles, minute gnats and leaf detritus. All morning this light shivers, wriggles, and scatters. Shapes stretch grotesquely, break up and re-form. Stress lines in the current stay firm, playing out from a boulder in midstream. All this light and movement open the endless variety of this landscape. My eyes find fresh new possibilities in this highway, in this fish house, in this refuse dump, in this body with long green hair, in this plow, in this liquid foot, and in this condensed rain. Each tea drop of this river is capable of bursting open and pouring forth a thousand new images.

When I drop into a rivervalley, I drop inside one very thin line, one private place. A plains rivervalley can appear so disregarded, looking somewhat like refuse next to the ordered lines of fields and roads. But I have spent one very rich time inside those cracks and tears, letting small events become woven into myself. I drop inside these lines named rivervalley, and I drop still further into small fissures that begin to expand into the infinity of smallness.

A river is full of small worlds. There are sections, quarter sections, acreages and small lots that all bring a good price for the thirsting mind that finds them. Each space is distinguished by a different play of light and color that might range from green or oily black-green to mud to bright-white. The surface of each lot has the undulations and flow of draped muscle. All this streamvalley and as far out from it as I would wish to go, from earth core to star and beyond, is in motion, streaming, and essentially a river of diversity. The small show me the way in which nothing is simply one event.

Gnats are tapping at the lid of each tea drop of river, picking up the dead. Jumping into the air in one place, they have to chase this moving drop downstream. Each tea drop of river is going by at perhaps two feet per second. Each runs up against a rock, spins sideways, and spirals off to the left and out into the mainstream. Imbedded in the surface of this tea drop, bubbles of foamy protein—complex magical molecules rubbed off the dead somewhere—are being tossed out of the tea.

Underneath this tea drop of river nothing stands still. Free-floating algae, plankton, and detritus rise and sink and rise again, roller-coasting downstream. And when this small landscape goes over a deep hole, the bottom drops out, nearly knocking it out of existence. This drop keeps a taste of its originality, but parts of other drops merge to spice the blend.

Despite this near disappearance, nothing is lost. Each tea drop still deepens the subtlety of the larger body of the river. And each drop brims with the river, and there is one tea drop after another to which I can turn time after time.

There is always a tiny sound, a minute cry, a miniature laugh, and a particulate sigh waiting to be heard, but often never noticed. And looking out across this rivervalley, every leaf edge, every underside and underbelly is stuffed to the gills with drama. The world is transfigured by the small and built from it. A small log is an immense highway for ants. And the interior of each ant is another path. Down, descending into the small, opens an abyss. Here is a galaxy, with just its minor brim sticking out in the sunlight of this morning. So that in descent into one tea drop, I ascend and outleap this river and the sea.

In one of these minute landscapes of river surface, there is a big expanse of sky. I am there too, looking up at myself. I see a large range of sky and I see myself in the small. I am next to a feather. All this sky is packed into one teaspoon of river and this one teaspoon is packed, in turn, in duplicate, into the very small holes in my eyes. And moment by moment some of the most paradoxical images are likely to be found there in the small. Just now there is a water strider skating in a cloud.

Ascending out of one of these tea drops, the local place can continue to open into a fluid world of interwoven events. Right in the midst of a galaxy of starlike drops in the river, there are large stones in midstream. Standing up against this stream day after day, they conjure up a strong sense of power, being like a feudal castle that is able to bend the river around it, or like a death's-head or the back of some large rooting beast. Yet for all their weight and size, they are not enough to strangle the throat. For all their size, they just make the river appear tooth-gapped. Just now a dragonfly approaches one stone. The image of a castle dissolves to a flier and its conning tower. When I want to name this stone and ask

"What is it?" I really need to ask "What isn't it?" Every event in the valley, whether it is one tea drop world or some object within my sense of scale such as a tree, is a stream of possibility. In the ascent of the day, when my eyes seem to be showing me the real world, I need to remember the biases of the eye and the wisdom of the fog.

With the coming of more light, the animals of the valley become chary. Inside the river, fish become shy. And overhead, a large red-shouldered hawk that is soaring above the river is circumspect. I am attentive, having been charged by the fog to look beyond appearances. Looking half-asleep, the fog has been a wake-up. It has restricted the landscape to shards, compared to the way in which the streamvalley is now suddenly all there in midmorning. And when the sun penetrates this streamvalley by midday, the landscape is so whitewashed as to be half lost again. Midday becomes too bright for comfort, which means that it is too far beyond my control. In all this glare, a fox could be standing right in the path of a hare, and a hawk unseen by the sparrow.

The lover of this light, the dancer that comes to life, is the plant, which nearly bursts in the increasing midday light. The plant has learned to moderate, to lean away or go limp or play possum in response to this awesome pressure. If my eyes are looking for a bloody battlefield, the noon sky appears to provide one, with vapor spewing out all over this streamvalley. But my eyes are being deceived if they find a war. Hit by this bomb billions of times, plants are breaking off pieces of this star and turning these waves into themselves. For all the power of the sun, this streamvalley brims over with plants. Not a war or even a relationship, plants and a star are more two sides of one coin, inseparable, like a river. A plant is a star in the earthstream.

The midday sun is turbulent, raging, inflaming, and vaporizing. The valley is blurred by bomb flash and fire storm and cannonage at noon, but the valley remains solidly present. It rides out a fire storm and, from it, builds a solid, tested foundation. Its plants have taught the macromolecule to live inside a fire storm and to take advantage of it. Their first attempts were near failures, and yet even the near failures, now dead and littering, have left a fortress of soil. And today, billions of years later, the battle has turned to a bout. And each new day is simply one more 10^5 second round.

Some days are less than blistering. The sun may barely put in an appearance. A cloud in the valley softens the sun. It is water piled up by the sun, waiting to go somewhere. A cloud is about the only way to get water out of the ultimate drain, the sea. Yet in trying to manipulate water, the sun plays a trick on itself. When the sun rolls this freight-train energy of rain-bound clouds over my head, they become enough of a force to slow a fusion bomb.

When a cloudy day turns to half cloud, the sky above the river goes from the appearance of a ceiling to one very deep river, an Amazon. On a half-cloud day, I am likely to see swallows hunting along the sheer white untouchable, tangled banks. I look straight up for miles. The sky contains a diversity that I readily accept in a river, but deny to that which looks like an empty blue hole. The composition of atmosphere is as complex as a river's tea. And when, for example, there does not appear to be a bird to be found in the rich woodlands, there always seems to be some bird in the sky.

By late afternoon one day on the river begins to cool. It takes a very thin yet broad form. Another almost imperceptible layer is forged and hammered into place. The stream-valley is becoming one very rich museum piece. If I would dig down a foot, I would uncover the record of a century. Tonight's flora and fauna will burnish the day's rough cast, making themselves in the process. The daycast reglazes the figure that holds the river. Here is one marvel that contains a marvel. Here is fine tea in a very fine bowl.

Something magical is always about to begin on a river. At dusk a half-burnt furnace shuts down for the day in every local place. A kiln mouth opens, and a bullion pours cleanly. Day is cast, lustering and crazed, congealing now into history. Just one spark drops, and that spark climbs upstream. Startled, a heron jumps out of the path of this red-hot slag, acting as if it had saved its beard of slender feathers. When that heron moves, I am moved by the delicacy of its flight against the rose-colored molten alchemy named Setting-Sun-Over-River. How can energy be made to stream in such a graceful two-pound form, heron? In that which had appeared to be a played-out, overused landscape, there is the magical finery that is a day, a delicacy and a power that matches the best of any world.

The half-burnt sun and its setting color on the river sink out of view. The colors of the forest seem to go the same way. Shapes wither and thin out. Yet even in ink, the valley holds onto a lively gray. As evening gradually swells into night, a sense of the familiar continues forward but slowly becomes something remembered. My eyes are getting lost, and something profoundly different rises out of the cast of the day. First there is a murmur, a humming that becomes thickly feathered in the cooling evening air. Smooth, gliding, hawkish, there is a dance of sound where there once was color. And in this music there is a remembrance of the force of all bells and their lifting.

Color stretches steel-wire thin to a salmon-tinged line on the horizon. It snaps and is gone for the night. Light has gone from a wheel by day to a wire above the terracing by evening. Then a switch is thrown and everything seems to grind to a halt. The plumbing almost stops in the plants. And yet the forest continues to be full of self-designing aqueducts, streamlets, and tricklings in the soil. By blind touch in the dark, this immense weaving of water seeps its way down to the river, dropping sand grains one by one down the hillside as if a hill were an intricate hourglass.

Above all this landed water, the darkening atmosphere also streams, weaving an intricate design in among leaves and stems. Below all this air, below the forest, the river lies so central, catching the last light of the day and the first star on its surface. Like the most consciously intended city, the river has chosen a route right through the center of everything. And yet a river is everywhere. The river cannot be said to stop at bankside, for bankside drips river that is falling off the hills, and it is also inside the sky reaching up over the river.

Deep below the comforting roof of the forest on this river's tangled bank, the land seems to gel up and stiffen. The day factory is shut down, cooling in the night. And the stars, moving at one-half billion miles per hour, appear fixed solidly in front of me. The river appears the same, looking like a black slate tabletop, even though it meanders beneath this bankside. The first movement of the night is sound. And the first sound I hear is the broad, delicate rustle of the river, a voice far spoken. This voice of the river is seldom heard in the day, and when heard, half-heard at best. All our attention is drawn in the daylight to its diamond-light movement. And at night, even right next to me, it almost escapes notice. It is something not made of words or even comfortable with words. But once heard, it is a finely tuned song that cannot wash out of my memory.

The river of soft sound is a bell. Its softness paradoxically strikes me. Suddenly I go from sitting by a river and watching it to being inside this sound, and I can feel, not just imagine, a streaming that is weaving inside me. Looking like nothingness, it is a journey both inside and beyond me. This *terra nova*—new land—of the local place is a courteous common ground between myself, a moon, a wave, and a molecule.

The deepening opaque night provokes clarity. Something is knocked loose by the bell sound. A crust is chiseled off, revealing more than appearances. It is a view that is no more or no less important than any other, simply strong and enchanting. And it is this wonder that draws me inside the dark.

In the swelling night, perhaps more than in the day, this river becomes a signpost with so many messages written within it as to be indecipherable. And yet the arabesque of its scripts, its fluidity, draws me with compelling force. Without my having to understand one thing, the river rains on my emotions, and within myself I hear a song of enchanted change in the way I perceive myself and my local place.

In the dark all these judgeless events—all greenery, star, river—are associated with the judgeful image of red bone-and-claw nature. There are voices in the night greenery, just behind my ear, that I elevate to terror, and my next step seems likely to be booby-trapped. This nocturnal landscape of hill and heavens and river seems capable of turning on me and breaking me apart, wounding, even capable of swift bloody massacre. I want to cage this night landscape. And I keep looking for a knife in the back, a viper's strike, a beast at the throat, or for a dozen powerful henchmen to appear out of the black wall of the forest or rise up out of the river.

Just as suddenly, when my senses tire of upholding this fear of the dark, I feel the opposite of terror. I drop my guard. Too peaceful a place, a garden emerges. It appears to be a place solely in my service, timeless, quiet, undisturbing and unchallenging. Over time I come to some balance. I see that I am a watcher and not the center of attention. I begin to become capable of being finely tuned. I put my ear to the ground. I move my house around and even begin to take down the walls.

The fortunes of plants and animals in the rivervalley seesaw through the night. Bird, in crow form, some twenty years old, is vaguely etched against the gnarled treetop. Claw is meeting clawlike branch. Below it, a mouse is foraging. In the distance an owl listens. Inside the river, a caddis fly nymph casts a silk net, hoping to make a profit in the night. In this hunting the valley appears brutal. Yet this rivervalley is kindness itself. It is wealthy and it is generous with its wealth.

The plants and animals give their precious wealth back to the world body. The river remains everyone's landscape. Washing away trails in sand and mud leaves no traces of ownership to delude anyone.

A gibbous midnight moon extricates my eyes from a back seat and entrances me. The spilt moon splatters molten silver on the river and sparks and spills again. There is a faint oscillation in the saturated air like a dream tone, accompanying this light. I presume it to be the hum of foraging midges. Off and on through the night, upriver, a dog's moon song can be heard. Pulsing wave after wave, molecules dissipate from the river, draining off of fish and mollusk, off decomposing plants and the leachings of meadows and rock faces. Without trying, light and sound and scent splice a loose terrain, an oasis for all thirst.

In the deepest part of the night, my eyes are still in touch with a star cluster that is light-years distant. For the most part, however, my open eyes hang like dead idols in a reliquary. I listen more than I see, to the river and to the holes that insects pash in this ink. Is it behind me or in front of me, their chant? One suddenly strikes my face, and there are others probing cozily at my wrist. And like the activity of insects, the trees seem to come together and press toward me. Wind from the hill, moonlight, river streaming, star and cricket are blended into a tea, and I adhere inside on a flavor— the river's tangled bank—sending out EKG and EEG melodies and chords beyond my perception. Soon I will be lost to sleep, beaten by the black sun, showing me in each day that I am inside and incapable of overriding even the night.

In the black night riverwood, there is not a green sign of life. In a score of different plumages, birds are perched and anchored. In the day past, they have cut this one streamvalley into a thousand domains and will likely do the same tomorrow. But in the night each bird has closed its eyes and chatters dreamingly. In the night it will make no life-and-death decisions. The streamvalley remains active, and yet the last service of the day, the compline, is one of peace. Moonlight dances off the branches in gulps, and scatters inside my eyes. This light colors a deep wilderness, bathing both this riverwood and my city. Inside this richly creamed tea, I am deeply home, satiated and inseparable and not having to go anywhere.

From just before dawn through the deepest part of the night, I have been inside a streaming that has washed gently and, at times, struck hard against my senses. Its turbulence has been enchanting and rock-firm: tucked and pleated, cobwebbed, waved and lacy, roaring, crashing and softly pushing up. One moment it is above and broad—the diamonds in the heavens, as vast as worlds. And then just as suddenly it is below and small—a river still, streaming around a stone in the earth, crawling in a wormy way, tendril-tongued and rooty-fingertipped, and rising up beside me under bark and wrapped in the embroidery of a leaf.

Without measures or even my affection, I come to spend a day and simply empty and behold. I give myself to the pace and the rhythm of this rivervalley or any landscape. The essence, rivering, is revealed not as a book of facts but, at best, as commentaries from listening points or visible currents. For everything I find to "explain the world," there is a counterbalance, which appears to be an opposite or at least something quite separate and distinct. For example, that which appears to be water, river, is a fire of energy in the cosmos.

A bird and a dog are distinct and yet the music of my river. Birdsong is perhaps best known because it takes the day shift. Dog has taken the night. A bird along my river will rise to a crescendo by midday, but a dog never does. By noon a river dog will pool up inside itself. And a bird chants with the clipped energy that sounds more like the sun than a river. A dog seems to be an instrument made of thick bone. It is not a reed flute by any means. It seldom speaks and is more likely to use its nose to lay a scent of conversation. But by midnight a dog is likely to sound along my river, banging against a mountainous black wall of scent. When you are living by a river, the song of one thick bony horn baying downriver is one of the most dependable musics of that river.

A dog's nocturnal baying along the river is perhaps also strongly our song. The soothing purl of the waters generates a fondness. It is perhaps a vague yet powerful remembrance of our rise to civilization on the alluvial plains of rivers—the Tigris and Euphrates, the Nile, and the Indus, to name a few. Lakes are rather short-lived in comparison with rivers, and fish might have languished and stumbled if not for the dependability of rivers, and so also, ourselves. Rivers have been our access to the waters. Our dependable routes homeward—our compass and our carrier—have been the rivers. Their waters read like road maps, with a precision that extends far beyond us. The salmon seem to be saying that each river has its own taste. We have lead dogs to this sweet elixir, this wake-up tea.

Here in this valley is my river, home, from-where-I-come. I listen to the wrinkles in my river, and they sound cranky and bent, local and peculiar. And yet the wrinkles in this river turn the sun, one yellow coal, into ten thousand suns—one star reflected in every ripple that turns the river in front of me into a long curving arm of a galaxy. The local mirrors the cosmic. This view engenders in me the recovery of a small degree of the river's profound restraint.

Against all my desires, it has been so easy to dismiss a river and waste it. At first glance a river appears inanimate and dull. And yet, when time is spent near a river, it begins to appear intricate and alive and inseparable. A river is an artery nurturing my life, and I am an aspect of it. I breathe and a new channel briefly opens in the world. A stream passes over bloody, rivery lips, then is dry on the teeth, and finally lost down a warming tube. This streaming within a river and within myself never ceases, never abandons, and can be depended upon day in and day out, night after night.

This river has an antlered bank of willow and cottonwood. In summer evenings katydids form a tin wall of music beyond these trees. In the winter a *CAW, Caw, caw, caaw* trails off downstream like a prayer wheel. The river is rolling, folding in upon itself like a whorled flowerhead. Here are the folds of a landscape that are beyond anything that I could imagine, full of the unexpected. Is this household or cathedral or somehow both? It seems to be arching into the sky along its bank, but it also appears rooted and grounded like a kiva. It is a living landscape that tells its rhythmic tales, a sand grain at a time and in no hurry. I am there *inside* that tale, perhaps for a brief flash of geo-time or perhaps as a vanguard of new opportunity and stability to come. I sit "sheltered" within the four walls of a cottage in the city, but my real house is the world itself, which ultimately sustains me. I am an expression of the landscape. When I am perched on a log over the river, wildness hangs over wilderness itself.

I keep attempting to sit down by this local stream and to take from it something that I can bring home to culture. And yet this rivervalley is my home and I do not take it anywhere. I want to think that I have surpassed nature, and especially this rivervalley which appears to be subject to my dominion. But when I cast a stone into this tea, I have no idea of what I have done (and I do much more than simply cast a stone). Has the world been reordered or is it essentially the same? If changes are to be made, I can confidently put more faith in the world, which knows how to make a steep vertical cliff and a flat riverbottom and how to fit a protozoan inside of a mosquito when I would not even begin to consider it. Take one fish from the river and the conditions for all fishes change as territories change. Pull one "weed" and the conditions for nearby plants change, spreading out like water ripples in an invisible yet resonating deep chord. Evaporation can take a landscape of microbes, and a rodent can take a grass plant, and a larger fish a small one. Commonplace pigweed, growing in the crack in the sidewalk, or perhaps the rare grass growing in a tropical forest, may save us all. The obscure events may be billions-years-old sustainers, acting masterfully without having to know the whole structure, being awake, digesting enzymes, for example, or seeing with bipolar color vision as I do.

I imagine myself to have won the gift of travel to anywhere that I choose. The last place, even now, even inside this miracle landscape of the rivervalley, would be to where I stand. And this choice is likely a measure of my limitations more than of my forward vision. Homeward is perhaps my most distant journey, veiled by the near-impenetrable subjectivity of my familiarity with this rivervalley. And homing is perhaps my most courageous exploration, because of the way that I come face-to-face with the bugbears of both the ambiguity of the landscape that I discover and its challenge to my most fundamental beliefs of separation and dominion. But without that homing sense, I might travel around the world and miss it. With my first perceptual step homeward, my fear melts to wonder, like hoarfrost or dew that dissipates instantly in the penetrating light of the morning star, the sun.

I am always in motion, in flight, even as I sit. Without taking a step, I am tugged and spun. My rivervalley is dragged along at a rate of at least .00011 centimeters per hour in continental drift. And this valley is whirled both at a thousand miles per hour as an aspect of planetary rotation and at 700,000 miles per hour as an aspect of the sun's galactic rotation. Feeling like stillness to me, I live inside a high-speed top, where dusk falls tens of millions of miles from dawn.

My journey homeward is more than a return to a place of birth or to a place of choice. Even if one's local place seems to be far from one's *sense* of home—being, for example, a harsh cityscape of exhaust fumes and bellowing sirens, or a country with a foreign tongue—there is a grand stream that overrides and interpenetrates all landscapes and turns them into home. My journey homeward is a revisioning of perception. If I will go back inside a recent special memory of yours and mine, if I will come from behind the moon and see the earth whole, I find the home to which I would gladly race. And if I was to fly down inside that landscape, I would find a feast of the most impossible intricacy, brimming over to the point of appearing thrown away. I might stop there and name my homestead Two Rivers or Blooming Prairie or Moscow or Paris, but its city limits would be as vast as my first glance of earthrise over moonscape. My homestead's essence would be with me always, and ever-expanding into the infinities of largeness and smallness. And I am not discovering it as much as discovering the way in which I rise from within it as its expression. I am a seed of this homeland. And I can fly out on the windstream and the pathway to other landscapes and bring them inside my identity because this home continues to swim inside me and keep me going.

Where are the boundaries of home? Home is far more a journey than a place. It is an ascendency, a flight to consciousness, a coming inside of, if you will, the depth of place and self. And when I am home, my journey really just begins. And to really come home, I go out.

A house is transformed to a home by the three graces: security, freedom, and identity. They are generated by a home front that extends far beyond the scope of my senses. It is a home front that cannot be windowed or walled, possessed or stolen. It is the real sense of home that can be within any house, that buoys me up, that frees me, and gives me that sense of from-where-I-come. And no matter what I do, this home remains inviolate. It has looked like an exploding star that forms the elements of this valley and like a sea that forms the limestone outcroppings along the river, and like the ice of glaciation that forms the cobble and sands the soil. Now, below this goat-hill prairie remnant, my home wears the skin of a river snaking through the roll of the low plains of the midcontinent.

Our hundred million small perceptions of home have kept this waning coal alive in all our hearts. The smell of home has been cultural tradition: my *rulle pulse,* and perhaps your *bûche de Noël,* or jalapeño peppers or collard greens or kassi mooli. And yet all this enriching diversity flies back in a quick geological time flash to a grand streaming. If anything is wrong with our sense of home, it is *not* its warmth. It is likely its narrowness and, as a consequence, its movement toward boundary. If I limit "self" to be synonymous with ego or organism, I am vulnerable and open to self-delusion. But there is a broader process of self that can be trusted. So too with "home," so that as I "expand my house," I fall inside the support that gives me the three graces. All my cottage and its backyard and the local neighborhood and its township and as far out from where I stand as I would wish to go, from earth core beneath to star core and beyond, all is essentially home. In this time and place the biosphere—that thin membrane of atmosphere and landform and ocean about the surface of the planet—is perhaps my vast immediate home, and yet it is still only one anteroom of one star, the sun.

From-where-I-come is more than a local place. The deeper body of my homeland is the river eternal. And no matter how far I wander, I am still inside this river as one of its forward-reaching currents. Whatever the crisis or the annoying distraction of the moment, a deep economy and a wonder remains at hand, and something toward which I can home. On a far-off bend of the Amazon or on the Moingona's bank beneath my goat-hill vista, in the dry bones of a desert river or in the straight runs of my city street, I find a down-home eternal music of diamond reflections on water and the sweet wind and the earth underfoot.

Home is a flexible sensibility that can narrow and then broaden. It can be reduced from a planetary consciousness to a nation or to a house in a township or even to the narrow terrain of oneself, markedly intensifying but increasingly vulnerable. Opening, home explodes to a disconcerting question at first, and then slows quickly to a wondrous exactitude. By its vastness, this grand household offers respite from the intensity and the blindness of self-absorption. Opening, I fall inside a household out of which I can never step. It is a house that can never be stolen, and that is my foundation. Where is it that a river runs? I breathe and it rushes inside my heart.

Our contemporary scientific measures reveal an uncertain and unbounded landscape. It is a vision that is both profoundly beautiful and perplexing. How to live—to navigate, to be home—in such a terrain? We have tried to remain islands, to keep our heads above water, high and dry—culture above nature. In dominion we lose compassion and wonder, and in our separation it is we who are locked on the outside. It is the thirst in our land of milk and honey.

How to live? There is, I believe, a way that is opened by the personal experience of a landscape. By relaxing and emptying, rather than by going with a looking glass or even our affection, we recover a sensibility that stands to complete us. Just beyond the window frame, in a terrain that is presumed to be asleep, everything is alive and communicates deeply: raindrops on the roof, weathered wood and rust-blistered metal, and the soft living resilience of the mosses. Every event opens to a river: forward-reaching, rapid, then slow, droughty, then a flood, beginning as an exacting thin trickle and gradually widening and dissolving inside one of its larger names, the sea. In the most obscure nuances of the everyday, in the tiny movements of ants or in the fluttering of leaves, I find a complex inflection of a deep, young poem, cosmos. Inside all things inseparable, outlands dissolve to interiors and I am deeply home.

Streaming

The sky is melting from high-white to the green underbelly of a forest. I have left the river's tangled bank, and I am gradually moving up a feeder stream of the river. Behind me, there is the fading, fluttering cut of the wind in the river willows. Ahead, there is the slow-draining voice of the rivulet, the soft clicking sound of a handful of pearls. Cottonwood has gone to basswood and black maple and hornbeam. There are red baneberry herbals replacing sedges, and gradually narrowing walls of sandstone, overlayed with crustose lichens and edged with slender cliff brake and liverworts. I have cast myself into a delicately tuned low light and quietness. My slow trek is an epiphany, a call to openness, a call to make visible.

I stop and balance on the creek's round, moist cobble. There are diamonds scattered sparingly on clear-as-glass shallows. I am trying to decipher its script of sand riffles and its punctuation of small stones and the random comma trails of fingernail clams. The sand reads like an automatic painting, too consistent to be disarray, and yet ambiguous enough to be patterned chaos.

A shoal of minnows flashes out of this pane into a deep pooling curve and is gone. A more minute, fine-glass shoal of spawn sparkle in erratic flight, afraid to fall into the dark. Unlike the oily rainbow twinkle of a turned lead glass goblet, these flashes are complexly alive with their own turning. Looking like romanticism, this view is the unabashed wonder of reality.

I have come into the heart of this dank hollow with too much sense of urgency, as if this place stood to be suddenly lost or as if too many moments had been picked from my pockets by petty demands. And I have been doggedly pursuing the essence of "streaming," as if it were a one-word answer. And so I yaw from my straight-line heading and take a deep breath and empty. My creased forehead smooths, like the surface of a windless river. My eyes dilate and dull from acute scalpels to broad dowsers. I pull the texture of this notch over my senses—shade, dampness, and viscid atmosphere. Suddenly there is a tiny spark of light in a sunbeam. It is a small flier zipping down to a fetid carrion flower. And when it kisses that flower, by this romance, a lost key of mine is recovered.

Inside this damp notch, a carrion flower runs away with my eyes. It might have been a coneflower on the goat-hill vista or any one of ten species emerging weekly in this rivervalley, from spring melt through the waning serotinal fire. The delicacy and power of the river and its forested hills shrink in the cosmos to an imperceptible dent on a speck of dust. And yet one flower can overfill my view. But this time is the one-time-too-many, and something snaps. In the flower there is a vagary, an extravagant question swelling like a seed, and it has the promise of becoming so much more than I could have anticipated.

The appearances of dawn or of my flower seem to offer more than enough. I want to keep this flower and all other attention grabbers. I am glad for this flower, for this dangling participle on a whip stem. And yet I also want the sentence—the season that has called forth the stamen and the petals, and the season that will carry forward in the seed. What draws me forward into the flower is more than its dazzling appearance. It is a sense of connection. I want to feel that which my reason has been telling me: reduce the flower to parts, and further, to cell and molecule, and DNA pops up to turn both my flower and myself into one diverse river of genetic gossip.

My carrion flower sings of connection. It forms a bursting green star. Within this flowerhead, the sun is becoming the future of a plant. The flowerhead looks like a whorl of un-curling smoke or a knot of landscape being untied. It is a music swollen to real-life honey, nectar. It is something white-hot inside a cosmic force, the evolution of a star, that is raging at a pitch-fire pace. Multifaceted eyes of small fliers push deep into this flowerhead. These small lives seem to be ca-pable of penetrating deeper than my eyes can take me. These insects go inside as if home, and even more, as if they were a flower. When one *completes* the other, there is no need to knock. Permission is granted.

Wherever I find myself, there is always a flower to steal my gaze. But center stage is not everything. All prima donnas dissolve and evolve. What blooms inside any flower is a fluid, eternal promise. Each flower is one fast pin of color in a season's parasol that opens in this rivervalley with five-pet-aled crowns and closes with sunburst composites. Inside each flowerhead, seeds ready to fly outward and split the boulder. The small seed can accomplish this, because, like the flower, each tiny seed is a channel for the flow of energy. What sustains through all this change is the stream that is in touch with all.

Against this pulse, I saunter forward and ascend a steep-ening incline. The sandstone walls narrow and twist until these stone outcroppings hang over my head and I am buried inside a tunnel of stone. My hands reach out and meet a three-hundred-million-year-old warm coal swamp now seized in stone. I touch the beginning of the Age of Reptiles, hundreds of millions of years before the emergence of my own species. As I begin to rocket through this time door, something immediate and local, something more than a flower, jumps out, like a finger pushed into my face. I have seen a bird, and it is like a hand held up in my face. A whole landscape may be streaming in, but nothing else can occupy my attention. Scarlet, bright blue, or leaf-tone bob on a branch. Their quick spurts are as attractive as a giggle in a forest, and there is simply no avoiding it.

A bird seems to be more complete than a flower—one solid color in a transitory landscape. Today it is a brilliant male cardinal kicking up the leaves in a small field of wild ginger, and yesterday it was a downy woodpecker overhead and a nuthatch walking down a tree. When I see these birds, it seems to turn coal swamps back to stone again. Appearances flower to complex islands, and they are more than enough. I know, for example, that the skull of a bird weighs about one-sixth as much as the skull of a comparably sized mammal, which both reduces weight for flight and cools the bird, with perhaps only one-fourth of the bird's respiratory intake used for breathing. And I know that the temperature of a bird functions to maximize food energy needed for flight, and that a bird defecates frequently because it has no bladder, which would add weight. And some of us have even taken the time to calculate the feathers on a chicken (8,300) and a swan (25,200, with four-fifths on the head and neck) and a hummingbird (940 feathers), and to examine an individual feather and discover a design consisting of more than a million components.

A bird on a branch is a complex island. But a bird falling off a branch becomes a process, a streaming of events. Billions of cells let go of the branch and take an earthworm, another billion cells. It is like a bee fitted to a flower, more than one life taking another life. And when a bird leaves its branch, a landscape goes with it—a score of twenty thousand species of feather-eating lice and mites and beetles, and worms and protozoa, and bacteria and viruses—and the very future of birds, bundled in an ovum or in sperm, and the living vestige of a dinosaur. An expression of the forest and a condensation of radiant energy fall off a branch and sweep from one side of the rivulet to the other.

When this bird sits down again, for me, it is no longer an object. It chatters like a river's rapids, and its heart beats like a river's runaway flood. And I no longer see a bird simply as an island of complexity, but I remember those moments when I held a bird in my hand and sensed its heartbeat. My bird's flight—its freedom—is an expression of rivering. And its chatter and full day-end song seem to be close to what I would say about the profound enchantment of this rivervalley.

I am attempting to walk on the thin ridge between the transitory world that I see in the everyday and the eternal world that I feel. My approach has been incantatory, going back again and again. Just now the sky sits down on a twig. It mashes and swallows a fly. It has a knifey mouth. A covey of earth rises up and surrounds this sky, waiting to steal any morsel. Inside this sky and this covey of earth, there is a rapid stream of rich red blood. I have learned to see the object first and foremost, and I seldom get beyond it. And yet there is a clear, obvious process that is a stream. Perhaps I can learn to bury the body for a while.

My flower and my bird—the life above the surface of my river and rivulet—lead irrefutably to a streaming just beneath the ice of their surfaces. But the surface of my river and rivulet is connection. In their surfaces all events, even my flower and bird, are reflected back as blends and amalgams and coalescent events. This weaving and eating—this surface-breaking—that the river does for my eyes is a way to recover the senses. And it is joy. Just now, arching over the water, the river grape gets leafed in gemstone inside the surface of the river. The sun and a woody vine are being romanced by the turbidity of the stream.

My prairie rivervalley is a fluid tapestry of current that gathers and transforms energy. Each appearance is a strand— a current—in a birthing and aging tapestry. My eyes touch a flower or a bird, and then I move off. A flower is going with its ax, rain, to the soil. Perhaps we both cross again soon. Perhaps one of us will wither at the next warp. Each strand threads in and through, up and down, touching countless other strands that thread and stream across it, all moving forward, weaving the appearance of a valley.

My flower and my bird—ground-rooted and high-flown— are exquisite pathways into the landscape. When the sun catches an edge of either event, I can begin to lose the body for a time in the scattering path of light off the object. That is what I follow now. One small pool of color washes through another larger pool. Iridescent waves break inside my eyes, and deep absorbent tones collect events together into small lakes of hue. In all these instances a thought goes out to the edge of my flower or my bird and it comes around again. A circuit in the valley is completed, the shuttle bar is pulled, and another thread is laid down on the loom.

A landscape is a tapestry, yes, but one too fluid to be picked up by the ends and be examined. My reaching arms become the landscape's currents rushing inside me, carrying in the irrefutable evidence of the waters, the streaming. The verdict is unanimous. From each event, a long curve sweeps out and melts deeply into time and space. And all the streaming of this river and rivulet—the flower and the bird—streams in my city's back street, down the main avenues, internally, in the vacant lot and abandoned house, in the kitchen and on and inside the powerline. One moment it might look like a withering, and in the next moment a swelling, then a bursting or a blossoming, or an oozing or the icy rock firm. There is a streaming, a rivering, and that is becoming obvious, and yet what is *that* essence?

How to grasp hold of a river? I was about to be privileged to be inside an event that could contain the bridge, a storm. From top to bottom, the rivervalley was about to open to the structure of a verbing, not only assailing, but also, upholding and affirming, applauding and bearing forth and conferring upon. The stream was about to flash-flood inside me, washing away the old question of "object" for a new one. From the commonplace to the rare, from the arc of a water ripple on the river to an orchid in this dank hollow, the streaming essence of the rivervalley would no longer just be seen but rain inside me, striking my emotions as softly as the texture of down from a hare's nest.

A few branches clattered. Maybe there was a need to alter plans for the day. A mouse nicked its eye. It had tried to sidestep that which was about to occur in this rivervalley. A hard rain came fast into this streamvalley, and it crashed as a boiling cauldron, spilling all its energy here and none of it there. It was refusing to go down a broad drain. Seemingly wasted in one place, the rain did battle, striking my face, then lying on the ground, looking like some fold come out of the sea, so misplaced. The rain did crash down, clinking and clattering, as one sharp presence of energy.

A wide-eyed fish whisked away. It was being buried in rain. And it knew that it was in too thick of an eddy of molecules, in too loud a chorale. The rain was upturning the fish's turf with a strange new pressure and streaming. Something was free and there was no telling what it might do. Some kind of mad body, something that smelled like everything that it had rubbed against, rain was taking over in the river, having cascaded out of, first, a meadow, then a forest along the river.

The rain washed out of these grassy and anting and treeing places. It was spitting and sputtering. And as it descended, the rain was sweeping legs and wings and grains with it. The rain reared up and over there and swerved here. It took one direction and then another. The rain appeared to turn on itself, suddenly banging into a stone, having no place to go, until time changed and the rest of its body caught up with it.

When rain meets up with wind, all lives had better watch out. A gale wind can punch and stomp. Tree branches and grass blades and flower petals and newspapers and cans and houses can be spun around, fall and shatter. The wind can tighten these events like springs. Broken, shouting, calling, the finery of a landscape is undone, but the rivervalley remains present. Jingling, eager, and pale, the wind is not all loss. Events are balancing. The storm is enemy to the drowning, but a kiss to the thirsty. A newspaper seems to lose its life and wraps around a tree, and yet new combinations of events are born.

At dawn the streamvalley had been rose-colored, then gold. The morning before the storm, the sky began to gray, with the cloud cover deepening and lowering and beginning to rotate across the span of the entire prairie. Then suddenly clouds changed to blue-black and this whirling wall came flying downstream into the valley. At first they entered as these hard rains. Then the winds, once only weavers braiding through the trees, were unslung. Full of matter, yelling and screaming, they pressed against a million covered eyes. And eyeless, a few lives ran in a right way and then in a wrong way, going head over heels at a corner. The winds had split to become a harsh smoke, an eye squinter, a blender of sound and smell and color, and a high-pitched tree singer. Storming, these winds jumped the dense hill and rammed into the open space above the river.

A storm passes rapidly. But the land that it leaves behind

is barely passable. The valley is ripped and torn. A flower
loses its head. Seeds eject. The ferns have been hit by a falling
tree branch and are broken. Roots are straining.

The storm has brought out courage, the kind of valor found
in battle. There are visible heroes of this storming—a Her-
culean root and an Amazonian mother—and there are also
the unsung lifesavers of this unrecorded battle. The roots, of
course, have held their breaths and saved just about everyone.
A bird has gone ahead of the storm as a messenger, and it
reads like an Extra or a Bulletin. A fish stands in the torrent,
hovering like a plate on edge, putting a quick end to those
who have fallen and are lost in the river, those who have no
reasonable means out of this torrent. Families have come
together and have saved the new chances, with parents turning themselves into roofs.

After the storm the valley is a landscape of bantam pools
and dips. There are a billion lagoons and bayous tucked away
in the grass. These reservoirs magnify each blade, so that the
smallest event is jeweled in praise as a survivor. The rivervalley is swollen, ready to burst like a droplet itself, heavy-
hung with all this water weight. The sky is all over the ground.
The passing wall of the storm is left in good measure in
thousands of pools, and I am also there in every one of them
into which I peer. Just after the storm, for all the disarray,
there is this wild, wet, crystalline moment, so pure and soon
gone.

Along with the brief rain pools, the rapids of the river
reveal anecdotes about the storm. They are like a word for
storms, and in this valley an epilogue. Like a storm, a rapids
is abrasive and spittled. There is the look of madness and
turmoil, and also delight. A rapids is a one-chance place, a
transitional zone, and a bowery. It can sound like a motor,
and only the rough and ready can afford to live there for
long. In a river, a rapids is a constant storm, sounding like
dropped dishes and silverware, like laughter and bantering,

and like a shivering and suckling. It is dinner for some and catastrophe for many. It is oxygenation for the whole of the river, and upset and muscle. I could not begin to imagine designing eyes and ears for life inside a torrent, and yet they are there in force. And to find so many different designs gives me a luxuriant bounty upon which to lean.

The stones of a rapids are astonishing events in their own right. Each stone manages to stay rather fixed in a topsy-turvy continual storm. Solid torsos, legs absent, blunt gluttons for punishment, these stones seem calm and enduring in a landscape of stress and pressure. A stone in the river is something upon which to hook both my eyes and my thoughts, like the anchoring stones of a traditional Zen sand garden. There is not a pale, uneducated face among the lot of the stones of a rapids. They are worn and graying. They look like fingertips through which the river flows. And when I put my ears on this hand, the fingertips change to teeth, chewing a rolling chocolate soil.

The valley after the storm is ashen and moonish. It is as if a huge fire burning somewhere close by were giving off smoke and soot that fumes up into the low atmosphere and obscures the sun. This sort of day steps inside a smudge pot for its color. The valley appears homogenous—a common room with a pallid yellow floor and a cluttered, dust-covered work table along the wall, the hills. The charcoal traces of roads, the emaciated strips of stubble fields in the bottom lands— all these events are like the atmosphere in which they are embedded. There is nothing crystal or marble, glass or aluminum. No bright silks. No sharp and distinct forms and no tight lines. Day appears as a smudged rubbing. It is limpid, soft and noncompetitive. Shadowed, in half-light, there is a beautiful sense of the whole rivervalley, not details. Trees are seen, but only the tag ends. There is a sense of everything being unified and hazed—an attic, a room at the end of a high, narrow, encased stairway, an ashen place full of venerable pieces of furniture with just enough tag ends showing under gray covering sheets to get a sense of each event.

Smells blossom in the nostrils, carried by the damp air. I smell the rooting earth, clearly. And I can sense the exposed damp-black tree bark and basal roots. The north-side lichens and mosses explode to light-green. Everything seems to be wet-burning, and my nose is taking in a sweet punk. And the wind in the high, wet leaves, feathers.

In this sort of ashen day my hands and eyes tend to stay close together. My finger touches a tree bud that my eyes detect. This bud has a feel as rich as fine leather-bound books and tightly woven Bokhara rugs. My fingertips sensitize, and the slight wind of this day becomes a cool, soft substance. I push a finger into this plushness, no longer emptiness. In all this grayness there is a wave of abundance opened by my combined senses. Hands, ears, and close-up eye contact turn this ashen day into an endless market for my senses. There is the aroma of refined teas. There are billion-years-old tested crafts and designs that please my eyes no matter how close the inspection. There are millions of leaves that have the texture of fine handmade rice papers, and there are soft and hard, extravagant musics.

As the day begins to age, the wind begins to build from the pressure of the sun on the vast landscape. But by dusk the wind pulls rapidly out of the forested hills, going fifty to one hundred feet up, and scraping against the treetops. Without the sun to fire it, the wind cannot even sweep through a small woodlot. And yet, it still contains a surprise. This wind washes against the canopy like the patterned surge of surf. Fifty feet up, branches crack against each other like breakers against a reef. Below, where I sit, on the forest floor, it is a still pool.

For some time after the storm, the flowers droop with water beads. Warmed to evaporation by petal tissue, the beads collapse to rivulets across the petal faces. Unburdened, the flowerheads repressurize and stand erect. All these flowers have appeared to be teared with elation. And now all this joy begins to run to the river.

A fish puts more muscle against the current. It senses that this expenditure might be worth the effort. Tussocks of grain with insects clinging to them are flying downstream. The fish hangs behind a stone, legless, with its nose wide open, and its ears, deep within its head, listening for high-pitched sound. It must tingle with a readiness that goes out to the tip of its tail. A blood relative tingles for sure, perched on the swollen bank of the river.

After the storm the river continues to build to a joyful dance. Duck wings whir in convoy overhead. The wind sputters. The surface of the river becomes part tea and part wind. The fish are ripples dancing inside this ripple. In a river that is headed south, the fish act like waving hands that are staying home and waving good-bye. The swelling river wriggles and dandies and twists like a cuffed magician, but always escapes.

The swelling river rolls and folds on itself like a seashell. The river has the ecstasy of dance in it, always. And after the storm the river becomes a whirling dervish that is praying with its movement. Mud-and-stick construction, brushy, congested: Try to take this dance apart and it is only more precise and indivisible. This river-after-the-storm is as heady as brandy and cream, that is disguised as less only by the meagerness of my perceptual capacity. This swollen river is an arabesque and pirouetting landscape, keyed up and feasting. And for all its good weight, it goes by me with the levity of a flute and the woody rake of a tambourine.

For all the rake and ramble of the storm's aftermath, there is a jeweled calm in this valley. Instead of running away, I have run *to* the storm, listening to it as if it were a call to orders. Not an answer, the storm and its aftermath have become a bridge forward, toward the calmed eye, which turns all storms and all places. And so I go forward, calmed, into the energy of the valley, not to be answered as much as simply for the joy of listening. I hear a language beyond words, a music and a mother tongue, that I cannot decipher, even though I, too, may be speaking it.

I continue to remain on the tangled bank of the river below my goat-hill vista. I am at the mouth of the rivulet where I began my quest for the essence of streaming. The rivulet speaks as it empties into the river. The sand bottom seems to perk up and wrinkle at the gossip that this rivulet is sharing. It is gibberish to me, and yet it is an entrancing foreign tongue. I have decided to remain by this obvious parting mouth, and to mark down the meter and the repetitions, looking for that first small word that will lead to another. This language of all waters seems as ambiguous as the shady edges of the two middle letters of any word. It has the rich look of flight, like arabesque. Suddenly the first small word rises rapidly like a body flying out of the river.

Just now there is a splash in midstream. Some body has cracked its head on this river's roof. There is a very handsome dance left in that spot, full of water rings going out. Something dropped back under the obscure veil of this river, but it did not get away without this breaking shout. It is a fish, a warmouth bass.

By the tangled bank of this river, one small fish rips away from the shallows, leaving a small ball of suspended sediment in its wake. A big fish glides closer like a drifting log. And suddenly it hits this little fish like a hammer. The smaller fish extends its spines, and yet it is lost, going headfirst, already in the stomach. These rock-colored lives, wall-eyed and looking like a river, have cut this stream up into territories like the birds of the forest above their bony heads.

Off duty, a fish anchors up against a stone. Overhead, wood planks, plastic wrappers, bottles, polyethylene foam, and cork drift by. A fish is an animal in a skintight exquisite suit. Its scales read like an obscure computer printout. Little landscapes develop under each scale. Parasites and wanderers choose their shelters from a hundred thousand sequins. Protozoa, leeches, platyhelminths, crustaceans, clam larvae, and mucus and odor form a precise landscape, perhaps offering the fish chemical communication between same species and escape from predation. However, all these micro-environments are ignored by two bug eyes that have interests elsewhere. These two eyes are looking for meat to hang out along a spine that looks like a picket fence. This style of backbone facilitates rapid growth. It was ready and waiting, already well-designed, to store meat two-hundred-million years before the emergence of the insect.

Near the bank small fish have formed a chorus. This shoal appears as solid as a ball, and remains intact even when passing through a wall of water reeds. The front fish serve as eyes while the center fish look for food, so that schooling or cooperative effort reduces stress and sustains each individual. When one looks down over the riverbank from above, small hatchlings are nearly translucent and the only hues they express are a water tone and a gold air bladder. Below them there are chubs that seem to fear the surface. The new-spawn appear so beautifully choreographed as they pass along the bank of the river, just below the river's surface. Suddenly they fall en masse toward the bank with all flanks flashing at once. They re-form near the bank and school away, going down blind alleys, but staying all the while near both the surface and the bank. So many as to be commonplace, there is sheer wonder for my eyes in the design of each precise wet suit and in the way that they are woven into one dance.

The light inside the river is green-algae opaque. It shimmers from the reflected crystals of plankton. Fish hang in clusters like soldiers at outpost. Big ones take the midstream position. All are peering into a veil that comes downstream—day after day of alert meditation, with olfaction and touch and eyes spread out like sensory weirs.

For an eye inside this river that is looking up at the sun,

the river may be a glittering dance, a landscape of sunbeams. This is more than water. Put a microscope on any drop of the river and I find something bristling, alive and ticking, in every micrometer. There is something lucid and wiry and something armored or balled up in itself. Fragments of the dead slide by alongside a newborn diatom—a microscopic and crystalline house of botany. If there is any sainted event in the whole cosmos, it is this stinking spittle, the inside of a river. When I put a hand beneath the river's surface, I am up to my wrist in the heart blood of the earth. Here is alchemy, every bit of this tea alive, with even the dead in the process of immediate transformation to something else.

To eyes immersed in culture, a river can appear to be a familiar and played-out landscape, army-green and regimented. And yet to eyes immersed in the river, the landscape is bent every which way, and there is bounty everywhere, and fertility beyond my suspicions. Even the eyes of a river are rich, being rainbow-faceted in a dragonfly and moonstone opal in a fish—gem-eyes in brown, muddy velvet cases. I am especially drawn to frog eyes, which are more visible and an alloy of both the forest and the river. In a frog's eye there is a small sun, a refined golden, molten iris, at which I can stand to stare. Pop-eyed and swollen, a frog eye is up to and almost over the tangled bank like the river itself. This frog eye senses movement and aids in swallowing a meal by retracting into the skull. There it is, around nearly every turn in this river, so forgotten, popping out watery and elastic. In just one summer a frog may recapitulate evolution, sauntering from an egg to a backbone.

The eyes of the river can open the possibility of a river's being something precious, because of their obvious gemstone colors. In just a quick glance at a frog or a dragonfly, I am surprised by the luxuriance of color that I wake to in such muddied surroundings, and a trigger is sprung. If I was to follow the life of the river over the course of a year, its more

commonplace animals—which are regarded as ugly and secondary—would open the wondrous complexity of the river.
An insect nymph of the river—caddis fly, hellgrammite (dob-
sonfly larvae), stone fly, mayfly, blackfly and dragonfly—can
move from a drab undersurface insect to, in the case of the
mayfly, an angelic gentleness at maturity. In its mature "spinner stage," the mayfly floats in the atmosphere over the river,
crystalline and feathery. It has no mouth and may exist for
only a few hours, seeking primarily to mate in flight. Near
death, this spinner is angelic, full of heavenly transparency.

Filling the same atmosphere over the river, there are the
marauding dragonflies and small green darners and needle-
thin bluets, all ever ready to fill the jaw. And below them,
the river's immature nymphs lie in ambush, and clearly, they
are no dancing crystalline naiads. The nymphs are army-green
and bivouacked under rocks. They are hooked, hinged, ar-
mor-plated, and roped in place, in a rough-edged, rock-firm
landscape. I turn a stone, and a patrol crawls away as if under
fire. With 90 percent of their lifespan rock-hidden, these
nymphs do not easily give up contact with this stone.

Insect life sustains because it is a primary consumer, scraping algae, sifting drift, and converting detritus. Unthinking,
insects are thoughtfully compact, putting their form into bits
and pieces, rather than building a dinosaur that is four stories
in height. Still, even in their smallness, these insects are one
vast precision event. They are everywhere in the rivervalley,
in the soil and the river, in the air and under the edge of a
leaf. And there are other invertebrates within and upon each
insect, as if the insect were itself a river. There are mites under
the wing and protozoa within. And while an insect can easily
be crushed, it is nearly invulnerable as a biological event. Just
now a water strider is skating upon the river on earth in the
cosmos. And in the fly and in the mosquito that I brush away
from my face along this obscure stream is an eternal spark
and an elusive philosopher's stone.

* * *

I root into the bank just below the surface of the river. There is a sense of both fantasy and fright. The middle riverbottom is a clean sand dune, surprisingly not murky. But the river edge is a bluish-black molten cool substance that I can cut like the finest butter. It feels like what I might expect a cloud to feel. It seems bottomless. It glistens, full of moisture. It is a false floor, a fantastic door. It contains the land of the lost, River Hades. Fish spawn and tree seeds are drowned and settled here. Both are swollen to bulbous form and ready to burst apart. This muddy edge is something that feels like nothing, and yet it feels that way because it is swollen and bloated, as full as a cosmos. Solid yet fluid, here is another representation of the streaming of nature.

I sink my hand into the mud, and it feels like very soft flesh, and I become a muckworm in this fat of the land. I get a sense of dissolution—the mucous, the viscid, the fetid, the heavy obscurity of the nameless. Here is the smell of sulfur, of sulfides transmigrating unwittingly to sulfates. How can there be even that much reordering in such an unkempt place?

The mud is an ash heap and a place that is fully alive. It is a landscape of trillions of microbes. And it is the very life of this mud, not mud itself, that is frightening. I fear that these microbes might get from hand to mouth and make me topple with fever, then to wither and become this formless mud. There are no bright feathers in this part of the landscape. Life here has a dull eye at best and is typically faceless and legless. Yet for all my sense of danger, there is something as rich as chocolate, and it is everlasting, a gathering of all, a cool lava.

There is one animal in this riverbottom landscape that is perhaps the biological consequence of the river. It is stone-shaped and stone-firm. It is eyeless in a landscape where an eye is not a necessity. Like a river, it is armless and headless. It is a clam, and it is as well-traveled as a fish and as colorfully named as rivers: Higgen's eye, dee-toes, muckets, pimple-backs, monkey-boots, elephant's eye, slop-buckets, and pink splitters. Their rich dark-blue to chocolate-brown colors with pearly interiors are an astounding find in something so over-looked. And when the shells are found on stony points in this river, they flash out like gemstones under silt.

Appearing fixed to one place, clams might produce three million embryos, glochidia, that lodge on the gills and fins of fish and then encase in fish tissue, with most eventually being taken by flatworms and other predators. A clam will process perhaps a thousand times more water than myself each day. And these clams inhabit the prairie-silt rivers in a way that is not matched anywhere else. They feast on the coolness and silt and glacial sands. The clams of the free streams tell a story about the essence of a free river, as well as other short, wondrous anecdotes. Their life on the bottom of the Mississippi River is one grand story. The clam may stand with the future of the river itself, and yet, because clams appear to be so obscure and unimportant, there is so little information on their lives.

The life of the clam *is* the freestream—migratory. And our cultural evolution has been nurtured by the freestream, par-ticularly by the vast alluvial plains, the enhanced water qual-ity, and the river as a trade route. Now, with a globally expanding population, we have expanded far beyond the modification of the alluvial plain to a more comprehensive modification of the entire river, squeezing the river into a thin line. In fact, my earliest memory of the river was really of a dam and my gathering of clam shells and overturning stones at its foot. Eventually, by treks into roadless stretches, across fields and through woodlands, I reached the remnant freestream, still a wilderness of sand-bottomed shallows and clear feeder streams and marshes.

Our interaction with rivers has become shortsighted rob-bery more than interplay. Our caustic interventions with rivers reflect avarice more than the diversity of both our experience toward and our feelings for rivers. We continue to approach rivers largely as if they were enemies that steal something from us. Rivers are barriers to be bridged—passed over—or channelized to drain fertile lowlands and deepen the river for navigation, or dammed to retain water and generate hydroelectric power and become our weaponry against the flood. All this problem-solving is ignorant and clumsy and overreactionary.

Channelization directs a river to be closer to that which we believe a river to be—simply, water. And yet, by channelization, the river becomes runaway water, silted and wearing the bottom right out of the river. Reclaimed lowlands silence the rich unspeaking complexity of the marsh, a national treasure stolen for a few. The percolating effect of the watershed as a whole is damaged, so that waste is concentrated rather than recycled. The water loss extends from the marsh far inland through the watershed, in a lowered water table. The increased water flow bleeds the watershed like a razored artery. On the Missouri River, for example, channelization has resulted in a river that is too fast to be navigated efficiently, and that now must be riprapped extensively and at a high public cost simply to "save" the river rather than provide any additional economic event. Vast wetlands are gone, and inland wells must reach deeper into the earth, and all waters are quick to leave.

To "manage" water, perhaps as many as two million dams have been constructed. They retain nutrients and concentrate them. Outflowing water comes from greater depths and is cooler than the water that it enters. Water flowing over a dam dissolves atmospheric nitrogen and supersaturates that water. Bacterial populations are increased, and the river is made chemically unusual, not just by the dam, but for long reaches of the river. Dams close the stream to fish and to larval clams, and the dams impact generally on a river for fifty miles or more.

The dam was my first way inside the life of the river. I went down inside the gap made by a concrete wall and arching steel floodgates, and I did the impossible. I stood *in front* of the river. Below the dam the animal life was concentrated, pent-up against the wall, with large splashes in the heavy turbulence and crayfish and frogs among the riprap. Topside, behind the wall, the pent-up current became fat and dropped whatever it was holding, forming a sediment sink. A small part of the water flew over the wall. Then and now it was a transparent waterwall, but now it also expresses a sense of terrorism, with water like a finger forced along the edge of an ax blade. Then and now the dam wore the appearance of stone more than a machine, of a ledge more than a wall, and it continues to resemble the music of a cascade. There are other positives to repay a fraction of the dam's cost to the river and, ultimately, to ourselves. There is an accelerated biological production, an enriched outflow of nutrient microorganisms, and reoxygenation for some distance below the dam.

While a dam can evoke enchantment, it cannot veil the cost of the dam to the freestream. For no cost, the unmanaged stream recycles water effectively in a slow, mixing run. And this one factor outweighs all other economic gains. The freestream would be the dream gift that our children's children might have wished from us. It would improve the water quality and open the marshes. The fresh, deeply economic freestream wilderness would be the pride of the future, and it would express the reality that we are not motivated by the profit motive alone. If our activism in behalf of the river were to consist of doing nothing but beholding and listening, we would outweigh, by mountainous leaps, all the short-run gains of the dam. And within our ignorance, we would begin to activate the power of our innocence.

Skipping down the river like a stone, a duck whirs bulletlike and straight-lined. A heron seems to pluck a harp slowly, pterodactyl-like, seemingly overbuilt. It folds its neck back and cuts its height in half and lifts freely over the tangled riverbank. It takes hope to keep moving, and flight is decided positivism. With its wings outspread, soft crystal feathers catch the wind. A neck is stuck out and taking a chance, in the way that a plant sends out a blind root to seek its fortune, and in the way that we reach out and come inside the wonder of our local place. And all these flights are more a joining than a freedom, a merging with rather than a separation.

The tangled bank of this river looks like a frozen wing, but when I come back to it in another season, I am always astonished to see how far it has flown in its dance with the river. Trees and grasses tilt toward the river, half-fallen, counterbalanced by roots that have to clench their grip hard to match the river. The tangled bank contains intricacies of force that make it a brilliant scarf to fit the river's neck.

Soft and slanted, the hill is the extended brim of the river. The tree species of these riverine hills extend high into the sky and their roots go to rock base. They form slanted cathedrals designed more for sitting than for walking. And in these cathedrals I sit between the river and the sky, lost under the canopy beneath goat-hill. The forest floor is open and perhaps even monotonous. Base-rock ledges push out at stress points, and small clear springs ooze from these rock faces. My hands lean against these rock faces for balance in this slanted land, and I am steadied and centered in a grace that flies far beyond this slanted hill of home.

In the humidity of the riverine forest, shelf fungi undulate in rows along a log, looking like a chorus. They have made a damp-tight grip looking like wood on wood. These fungi have something direct to say about life and death. The message is beautiful and aged. Ashen, pale yellow, or mawkish orange, here are the flag colors of death—the warm and hot sleeping tones. Touch them and they bruise and topple as if they were always blood-drained. Their flowering dance is brief, snaking out overnight like an adder or oozing pustulant out of a log crevice. Directly linked to the dead and becoming quick-dead themselves, as well as being death itself in a few species, they are an awesome connector between life and death. They appear to be humility itself, staying with the undergrowth, but near the heart of the forest.

Light emerald ferns flourish along a descending wellspring runoff. And like this spring, the ferns cascade in trickles and splash down this rocky cathedral. Neighboring liverworts overlap each other like lizards stretched tight as drying rawhide on moist boulders. And there are other stones, skinned as soft as slices of milk-soaked bread with mosses.

One keeper of this cathedral, a grackle, waddles over the ridge of a log with its tail cocked to the left. And there is one fallen tree after another, just about as large as death can be. To an urban eye that sees a landscape of parts, these logs might seem to be just so many broken chairs and pieces of china. They appear to be refuse, decaying and rusting. And yet these logs are a house of life, with rooting and hatching going on inside. And when these logs are broken to their finest pieces, the myriad pieces continue to live, becoming soil and then flora and fauna. Molecules and ions continue to rub snugly up against each piece, as if each chip were a forest, and energy continues to stream through it as if this chip were an opening in the dam.

The light in this cathedral next to the river often diminishes to next to nothing, going from perhaps 5,500 footcandles raining down on the open river to 500 footcandles under canopy and down to two footcandles under low herbals. There are always small clearings to be found, islands of light termed sun-fleck illumination. The sun is ordered here, and I can get inside it and then out again. I can walk around inside this aspect of the sun and the forest, and survey the ligaments. It is like a brightly lit monk's cell.

Descending the hill to the brim again, I pass out of the

riverine forest through the willow and water dock and beg-gar-tick to mud-beetle silt that is seeped with river. Tracks of mink and raccoon weave this mud back up the hill. A flowering ridge of grasses edges the river, and I am out in the sun, drenched in its unblemished light. This hot open space is a miracle. Just the capacity to sense this light, a most common event, is magic. How can something so fast, moving at a velocity of 185,282.3959 miles per second, be caught by a sense? And mud-cracked and often eroded, this tangled bank is a fine stage for a miracle when I take a moment to consider it. This brim of the river smells mint-warm under the white-hot light of summer. There are white and pink blossoms that are like the soft lip of a cup, with a muddy tea swirling snugly against it. A goldfinch sparks through this lip, and a cliff swallow jets just over the topmost blossoms.

In the tinder grass of this brim, ten mice ball up. Two have long whiskers that go over and over eight pinkish baby faces. Blind, mouse babies have found themselves in one ticklish land. Grass and seed come spewing out of eight teats; mouths on teats, naked meeting naked, a channel is opened in the grand river of nature.

By this riverbank below the goat-hill vista, the river has widened to become a broad channel. The width of this open space draws the wind like a path of least resistance, full of river fragrance. Below it, the tea-brown river passes by, going in another direction. Black damselflies stay nearby and even go north against the current. My thoughts are flowing, and it becomes over-apparent that like the wind and the river and the life of this valley, I also stream in order to keep myself paradoxically together.

Just above the river a leaf drops. It spins down to the brim grass trailing a glittering mirage of color and looking like a wing. The wind picks this leaf up and takes it off to where-abouts unknown. Chance and order are at work. For a season this leaf had resembled a hand. It had been just about that size. Now, losing its lifeblood, it has become a shell, a brown suit and a hat. More and more this leaf is beginning to look like a brown tattered wing. Smoldering, rolling in the wind, this leaf sets down on the river. It is carried downstream and then blown off the river and wedged in between some stones. One spirited leaf thus stops, full of meat, and full of a season's stories for the soil.

This leaf is not weeping. It has swung in the wind like a gaucho in the saddle. And when it goes to the soil, this leaf does not cease to exist. It is still a fire. Color burns from it, and this leaf turns to ocher and then to a grayish cocoa. It smells more like cool tea than like death. A lot of autumn's unthinking life will have enough wisdom to weave themselves into this fire. And with the coming of ice, larger, meaty flesh will also gel up between these sheets, so that just prior to the first snow, this leaf fall will become a winter refuge and a glutton's feast for some of the wildest and most obscure events of the earth. There are, for example, the most common animals on all lands, and the most unnoticed, the mites and the springtails, forming assemblages of perhaps 25,000 and 10,000 animals respectively in each fertile ounce of soil. And there is the exotic diversity of the beetles, in their porcelain coats, glass-smooth across the back and abrasive and finely haired near the head. The leaf fall is a streaming tapestry, with every touch sensation present, from smooth glass and broken glass to bead and soft down.

One small leaf can represent so many currents. For example, a part of the leaf may be back under tree bark as an egg. A piece of many waxy leaves is lost to a gray caterpillar's bite. The tree responds by cutting off the water, squints and blinks, and cauterizes the remaining leaf edge with radiant energy. And yet a part of the leaf has gotten away, and it is soon to become a moth.

It is some wizardry that has stitched this rivervalley together. Loose cord, plant, is cast branch to branch and inside the soil. The plants have worked themselves down some treacherous paths to get to this valley—through drowning tides, along with the man-swarm of immigration, and butting up against glacial ice. As seeds, plants have become capable of being scorched by fire, passing the stomach, and piled in damp heaps and frozen. And where my bones seem to stiffen in the damp cracks and crevices, plants only become more flexible and wake to a cool fire. The way that plants will go to the landscapes that I want to disregard as a wasteland makes me begin to reconsider my view to be prejudice. If they will give their lives to these places, perhaps I should be coming here to do my fortune hunting, and perhaps *any* local place anywhere on the earth contains a diamond mine.

Each event seems to contain everything. There is, for example, the fish hung in midstream like a teakettle over a fire. Stuck in the very heart of a plains river, a fish is the river's way of breaking an insect, of how to remain cold in a cold, of how to do without a clear voice, and how to keep from standing out in a crowd. These ways seem to be the minor ways, but the minor ways of humility and restraint look to me, more and more, to be at the heart of the earth. A fish can be a door to the essence of a watershed way of acting. It doesn't seem to know where it ends and a river begins. But I am beginning to see that there is a fish in the tree and another that is spewing out of the sun. I feel the twenty-eight bones of my face and the twenty-eight bones of a drying fish skull on a stony point of the river, and I find the way that a fish becomes me, continuing to stream inside me, and, importantly, weaving me to the same overriding processes as the fish in mid-river.

I lean upon the dependable presence of the river's flow. What is it upon which I am leaning? What is the essence of all streaming? A leaf pirouettes and drops down to the tangled bank and then lifts and glides down to the river and sails downstream on the river's surface. I make my best guess as to which way it will go. It turns the other way. I pick out another leaf swimming right in behind it, and that leaf finds still another way to fly forward. Puzzled by this unpredictability, I look up to the dependable ocean of atmosphere, and, in the particles swimming in the backlight, I encounter this same irregularity. Upon closer inspection, the surface of a smooth stone is dappled and dinted. As I turn to the structure of a bank and also imagine the whole stream from a DC-10's altitude, all appears fractured, fractual. I am no longer simply puzzled; I am disconcerted. I have gone out seeking the fundamental ground of my being—the streaming—and it seems to be disarray.

What will the weather be next? There is this built-in chaos in all landscapes that veils such forecasts forever. There is no longer any possibility of an accurate forecast of next month's weather. If the prediction comes true, it is a fortunate guess. Weather will always be a horse race. No more than a few days can be forecast with accuracy, because small weather irregularities—too diverse a sum to monitor with the most sophisticated cybernetics—become the vast weathers. Yes, there is chaos in my rivervalley, but when I agree to face it, I may find that it is not disorder, and perhaps the essence toward which I have been trying to move.

I have come to the rivulet and to its river for something more than appearances. I have found the stream of nature and its structure, chaos. And that which was, at first, disconcerting becomes rivers of exciting possibility.

The craziest implication of the unpredictability that is generated by chaos is that I am the better for it. First, I understand why there cannot be an answer. The discontinuity of turbulence in this prairie river and the valley's sky and the zigzag fractal geometry apparent in the hills and tangled bank assure for a landscape that is not predestined to evolve in a certain manner. Instead of a valley "on evolutionary schedule," chaos reveals a valley still inside creation. Rules are obeyed but not necessarily repeated, and the grand patterns of weather and seasons are aperiodic at best. The forest meets a different spring. The discontinuity that I name *chaos* is not the disarray and randomness that is popularly associated with the concept. Chaos involves the complexity of simple rules and makes events the better for it. For example, with a fractal—jagged and irregular—geometry instead of a smooth geometry, more oxygen can be absorbed by less lung tissue and blood vessels can access all body cells and still not exceed more than 5 percent of body composition. The nature of chaos affords opportunity, not loss.

I went out seeking a landscape of waters, a streaming structure. I have been seeking the eloquent smooth curves of a Leonardo da Vinci drawing of water as a reality. And yet, inside the chaos of this rivervalley, I find a landscape that is ahead of my vision. I begin to face my grandest fear—that, for all my education and my experience, I live in an ambiguous, unanswerable context, and that I understand next to nothing. I have found the waters, but I am inside a turbulence where all events, no matter how rock-firm, are chaotic waters—percolating, eroding, and tortuous.

Still, in chaos I can see a better landscape than the one I have been seeking. It is a valley with pattern, but open and wild and creating. At first glance this reenchantment appears to be a romantic delusion, but it becomes the most realistic, practical vision. When I go back to any one event in this valley, such as a bird, the view remains chaos. Even in a flock, flight is not robotic but leaderless, free and wild and chanced. Flight is order and it is creation on a wing. My consciousness melds, as I am biologically joined, to the streaming. My thinking becomes contemplative. I am moving forward, deepening the complexity of the rivervalley, alert and taking passage.

Overhead, the soaring hawk, and below, the stilled stone

are rotating like capstans, hauling in current lines, coiling them up on my pilgrim's ship, the river. Their glistening attraction is not their appearance as much as their vitality. Circling flight and subtle stone coloration are the turnings of the current lines, being wound up and let out into the main channel. By its activity each appearance becomes a precious moiré, an unending silk fabric engraved with a wave pattern. The red-shouldered hawk above the rivervalley, a carrion flower, a fallen leaf, and an ant become labyrinths lost inside labyrinths. There is a pattern, yes, but indecipherable, chaos. The task and the joy are to be awake inside this complexity, and to journey and behold rather than search for a way out. The journey's end is to open, so that never again could the sunset be dismissed or subsumed; never again could a bird or the grasses be demeaned. Each feathery barb and each glistening blade is resplendent.

Passage

Snowfall arrives. It is dusk. Upon seeing this soft wall roll over the hilltop, I rise for a better view. I stand on the river itself, on one diamond-glittering, marble-hard floor. The heat of the day is now only an afterglow of crimson and violet on the western horizon. And now this snowfall begins to erase this color. The river ice almost snaps from light rose to sea blue.

The snow descends quietly, muffled. It sizzles like dry ice as it meets my coat. It begins to erase my footsteps and round my shoulders, turning me into a small hillock on this glory road, river. I am wrapped up in a ball of down, inside this soft cloud of flowing ice. The wind-scooped snowdrift is sequined and phosphorescent. Ice-cold dry air is spiraling down and out of the jet stream. Full of ionized oxygen, it rolls slowly down the river. I am charged awake.

I savor the exquisite jeweling of frost that rises in small clusters over all the surface of the river ice. And there is the precision thrown-away luxury of a snowflake everywhere. Each hexagon may be easily derived from a million ice particles and from trillions of water molecules, and there may be a trillion snowflakes in a snowfall. And yet the overriding sense of the evening is not of these descants. The river and the hills come together as one clean page, as one whole body of melody.

The snow builds to stacks of irrefutable evidence for a new verdict of self and place. The snow continues to harbor a narrow, three-dimensional landscape of length and width and height, and yet even the physical form of the snowdrift reveals a more profound dimension. There is the presence of sublime wind-formed curves that run in long lines for the length of the valley. They erase differences and wear the appearance of currents, and they build to one grand sum, the stream. In each snow curve, there is the activity of the whole. The snow's arrival is the consequence of a planetary wobble. Winter is simply the regular sway of this rivervalley away from the sun. It has left the light slanted and shortened the day. But with another dependable wobble, winter will change to spring melt.

Perhaps more than any other season, sparse winter details point the finger of guilt toward grand processes. Lacking the rich variety of the summergreen rivervalley, the paucity of detail creates a sensation of wholes. With the details of winter taking a backseat, there is the desert's clarity of unison. And in the similitude between events, the remarkable grace to be experienced is the fluidity of time.

Yesterday, from perhaps someone else's local place, is in the snow's moisture. And I am inside a landscape that has become a small ice age, and there is a taste of having reached back into a time when the physical earth had its hour of triumph. The sense of the archaic, the ancestral, is taken inside my reflection. Time is less compressed, and it is remarkably more than this unfolding. The past is sustained, deliciously alive, in the immediate moment. Far away in Greenland's unfading ice, there is the fossilized record of sycamore trees that flourished there in warmer landscapes one hundred million years past. Now the sycamore is here, asleep but ready for spring, one hundred feet away. It is a light-colored skeleton against the graying sky.

Miles-high glacial ice lobes were integral to the formation of this valley. This glaciation can be sensed, of course, in the geological record of this valley, but it comes alive viscerally in this evening snowfall. I may have ascended out of a stone age, but I am still inside an ice age. I am standing on an ice floor, and time is jumping inside my senses in quantum gulps. Prairies that stood next to this rivervalley not long ago snap to the grain fields of the present. There are the limestone cliffs that rocket me back to the time of the vast inland seas, replete with corals. And what all it leads to is not simply elation, but an expansion of self and place.

There is nothing lasting a century. Nothing lasting one second. One micro-second past is history. My body has a new cell combination, and the atmosphere is a different mix of molecules. And beneath me there is a new river, one that was upstream and around the bend just a few minutes ago. Every event that I sense has changed, and, thus, in a very real way, is change itself. I blink an eye or take a step, and this valley has mixed to new color and sound. The stone that appears inviolable is being eaten and moving. The essence of this fluidity of time: The valley is in passage.

The predominant ambience of deepening winter seems to be a sense of the past, and yet there is always a subtle sense of the unfinished. My eyes are the prime deceiver. The valley comes in as history at any time that I look at it. Appearing fresh as the moment, my perception is always past. Overhead, the evening's fresh starlight is older than several of my generations that preceded me. And when I first sensed that I grasp nothing but the past, it was astonishing to look at the moment and look backward. This river is past; the star is past; everything is history. Made. Done. I blink again in my astonishment, and everything remains history. Even the ancestral seabed outcropping is one microsecond older. While astonishing, such a perception facilitates a view beyond the limits of time and the narrow three dimensions of space.

Every event is seen as done, as history, and yet each is far from finished. Each footstep and each glance in this valley is off a dazzling razor's edge. I look to the past and, in the same moment, I fall into the future. Each step is into the familiarity of the valley, and also into the midst of a grand abyss of space and time. I am coming to a sense of the ambiguity of this view as a privilege and not a burden. This view is an opportunity for freedom. I am adrift more than fixed. The uncertainty of my view is still far more than enough, and perhaps more enchanting because of my limitations. Because of the obscurity, young Jupiter twinkles like a cricket in the night sky.

My returns to the river in any of its seasons are an anchorage to a quest more than to a place. There is a dependable profusion of wildflowers, but that to which I cast anchor is the process that underpins and expresses that profusion. And that which is permanent, regardless of the degree of natural catastrophe or the degree of our degradation of the landform, is also in your local place. Like a drifting plankton in the river, my anchorage is passage. I fix my heading on the visible currents, the activity of plants and birds and rock and soil. I have no other practical choice. Over geological time I have observed this valley to be a quick-change chameleon that cannot be one landform forever. The essential permanence of all landscapes is a stream of change.

The river is my pilgrim's ship. Powerfully built and forward-running, a river is like an arrow in flight. Bedrock minerals, soil, new hydrocarbons and radionucleotides, crayfish and snail shell fragments, killdeer feathers, sponges and diatoms, leaves and polyethylene bottles are all on board, everything that is occurring in the world. They are all taking passage in both turbulent and laminar flow, and I am joined with them. This arrow, the river, is flying straight to the heart of my senses.

The rivervalley begins its sunrise meal of radiant energy. The sound of the river slithers by like a fish belly. I can saunter up alongside this vital stream, touch it, and wade in it. Hour upon hour, under the moon and under the sun, the river streams forward. It is surprising that the local place doesn't run out of water. But this channel is just the visible tip of the iceberg. This river is the sea, looking like a river to my eyes, and it extends much farther to be the swirling energy of the evolution of one star. If I could see directly the process that is occurring in this valley, it would be this river and landform and the sea and the sun as one diverse replete stream. A river is just a more obvious way inside the grand stream of nature. A river is so full of room that my body can sink right inside it.

Your local place contains this grand stream. It may be a path that is more condensed, such as a mountain range, or more sparse, like a high-plains desert, or even more open like an ocean of sky or water. It might even wear the appearance of something that first appears to be nearly the antithesis of a river, perhaps an inner city, which rages on this planet with the fury of a wild, young mountain river. The stream is still there.

The river in this valley often seems on the verge of snapping under the strain of so much liquid. In one small spot it can be bent around a rock. And yet, for all the shaking and rocking, the river is an endless force in its essence. I observe only a small part of it, and yet I know how far beyond my immediate senses it can reach. No matter how far out of this valley I seem to fly at times, I am always at the head or the back of the tail of its stream. In this valley I watch the particles of foam and spray being lifted into the atmosphere and appearing lost from the body. They rise to become the seed for a cloud, and they remain the obvious expressions of a grand stream of water going down the drain and coming around again.

While other landscapes can appear to be quite different from a river, all landscapes are come-arounds. When I return to the city or travel to a distant landscape, I look for a physical river or the dry bones of a river or the fluid roll of a city hill. In my stolen moments I anchor myself to it. I ride with it under the passing clouds. And I look for signs of its ancestry—stone, cobble—and behold its details, and try to grasp a sense of its infinities—the panoramic and the minute, stars overhead and the vista and the colors of its plants and the movements of its insects and birds. Like fire before my hands, this elemental curve of landscape can recover a passion in my everyday and, thereby, quench a thirst.

It is sunrise. There is a warm summer's rain that turns the sandstones to a tropical cinnamon. Created last night, a labyrinth of Oriental silk hangs jeweled on wood nettle. A starling's song calls like a muezzin, a prayer call from the high minaret. I try to listen to what my feet have to say and what my lungs are whispering in their small talk with the wind. They seem to be more in touch with the landscape than my eyes.

I had been to this valley at dawn perhaps a month ago. The nettles had been tiny sprouts underneath a wetland drenched yellow with marsh marigolds. The toads and chorus frogs had come together to sing. Perhaps five of each hundred seeds had germinated. The valley was tossing off its shells and seed coats and sheaths and snapping out of hibernation. There were new buds and insect larvae and eggs. The ferns were beginning to uncurl, and pungent aromas of opening soil swept in on the soft breeze. There were holes everywhere in the hills, full of fresh, rapidly beating hearts. Nothing could stop this spring or the evocation of the beginning of the summer fire.

A burning star condenses to a plant. And burning-star-to-plant is really the history of the earth and its current activity, including ourselves, in a phrase.

I have been walking all over this story of the earth. It is stacked tenderly under my shoes, and it is being lifted into the wind. It is not an old volume on a shelf, but a story with a nerve end.

The grand story of burning-star-to-plant is carrying forth in curious nuances of the valley. Surprisingly, it is a global story. There is an African weaver, a sparrow, and an Oriental carp, and a Russian mouse. There is the immigrant Queen Anne's lace. There are winds from China and water from all seas and all rivers. This tea-colored river and its valley, with its dead-leaf-toned sparrows and European flowers in the floodplains, are a league of nations.

It is difficult to book passage, to travel with any depth, when I am always on the move, either physically or perceptually. I not only can travel a good deal in one place, but it is likely the only place wherein I really begin to travel. I have to stop really to take passage. I see the stream because it pushes against me when I stop. When I sit down in one place, the past falls into the future, trailing out like a wake around a rock going downstream.

I might spend time with camels, or on a mountaintop, under sea, or fly across great fields of grain, or meander through the towers of great cities. I might see Tunisia, Florence, the Great Wall, Rio, or Paris. I might take a villa by the sea, visit mosques and cabarets, the English gardens, the ruins of Mesa Verde, or the cave of Lascaux. There is nothing wrong with this wandering. But there are profoundly deep experiences right where I am at any moment. An ice-cold wind, a quiet wall of stone, the spiral of a thousand plants in the river lowlands, or a shell's graceful curve: They are the real cues, a real tapestry, and a journey that is going on right next to me and within me. It is better than magic; it can be sensed.

All about me and within my own activity, I find nothing but affluence, running over the brim. It is an immensely real departure, uncluttered, beginning either with or without me. Instead of adding on, I can strip down, cast off excess baggage, and get down to the senses. The sun is shining. It is a fusion bomb. There is a wind, and it meanders everywhere and points the way—a watershed way. The wind meanders across a field like a river and goes over a road and a house and right through a chain-link fence. It is touching everything. It seems to be made of nothing, and yet it is moment-by-moment life to me.

Walking beneath goat-hill vista, I try to wallow, to stick my senses into everything. I follow the deepening grasses and the basin of a sandstone hollow and a million stones along the low water shore of the river. I try to walk blind to that which I have learned in the past, and to open to changes in light and texture and color. I also try to walk in sound, to go with meter of the hills and the wind. This form of walking will not really take me anywhere or come up with any solutions. Not being on a journey to anyplace or having to arrive on time, I am in the center of the stream.

Walking along the river, backwoods and off-road, I am purposely taking the long cut. I labor through dense brush and against the steep slant of the hill. I try to reduce my intent, and not to be in such a hurry to step into the future that I sidestep the present. Then each step, and even each pause, and each quick glance that opens to a long gaze is ascent and flight. I may walk over the terrain, but I fly within a great stream of energy. No longer sleepwalking down the straight and narrow, I am awake and alert and in contact with the elements.

By my long-cutting I find a shortcut to the landscape. Roads are strewn with ups and downs, bent and twisted. They can rock me like a storm, as they sidestep and try to go around a bubbling cauldron of energy. And where they turn or descend is where I want to get off, so as not to stay like a dog tied to a leash. I do not have to forge ahead and probe to find the land. I run straight into it. I take the road to where it turns, leave it, and abruptly come face-to-face with a torrent of energy.

Walking by a river is like walking before a deep cut in the veil, and seeing, in that cut, the deep stream of nature itself. And I can take my rivery view back inside my everyday and find the grand streaming that is still billions of years in the making. Smack in the middle of the landscape of a neighborhood or in details such as an electric wire, natural music spews forth still white-hot.

On this river's tangled bank or on a city sidewalk, I begin to use my dominant eyes in a more penetrating way. My eyes have been sealed up in my head and used sparingly, as razors, to cut the landscape apart or to plug a gap in the wall of my thinking. And yet my eyes can be open portals, and they see more than they want me to know. They can look through a pinhole and see a giant star, and they can jump time and space. They can find the cues that reveal a bird to be one suit that a dinosaur wears today to continue to exist. My eyes can see a bird eating bugs, in color. They can see a bird's knock on wood. Because of these images, imagine my possibilities. In 150 million years, I could be a flying elf, perched on dry twig legs like these saurian birds. My eyes know too well that anything can happen and that it will.

Everywhere there are little currents named light that speed into my eyes in millionths-of-an-inch-long waves. And along with my eyes, my blunt tongue and my thin nose hang like fish in the river, skimming cues from a stream of molecules. One small cue leads to another, and soon I recognize that it is not bits and pieces that constitute the valley, but rather the whole valley *and* cosmos that I am letting parsimoniously inside in bits. It is natural for my eyes to lock in on molecules and minute bolts of color, to assist in the selection of resources to satisfy a momentary physiological or acquired desire. And

I am not alone in the provincialism of my attention. A particular species of warbler may feed in the middle and ends of the upper branches while another species may prefer the base of the lower branches. A moth may seek only a particular flower head. However, it is unnatural for me only to spend my senses like a miser. Despite visual dominance and general sensory limitations, I can also hear an ocean's roar in a wormhole.

With the demise of winter, the valley seems to flood my senses with incoming stimuli. And yet, as spring flies to high summer, I sometimes seem to find less and less. There are phases in any season, and in any day, for that matter, when the valley grays in my senses and is temporarily lost. In these moments I find it to be beneficial to stay with my senses. When the landscape dulls, I find that I am likely to be on the verge of stumbling face-first into a new view. When details vanish, the grand processes may coalesce.

When I go rivering in the valley, I find water in the eye of a heron and in a violet's petal. With this sort of perception, I am not seeking an answer as much as I am trying to keep a door open, to remain awake. I am going after my neglected contours that are unbounded, uncategorized, untamed, and yet as real as a seed or a drop of water. This sense of direction—this passage-taking, this rivering—is my pilgrim's ship's—the river's—jetty. I can go inside my everyday with a recovered sensibility that does not codify, that makes transparent and unbinds or liberates.

Wildflowers appear glazed, like wide-open eyes in too much drought. There are towers of olive ragweed. Sun-bleached lichens on stones and trees curl on their edges and flake off. Parched pheasants shuttle through the tinder grass. Above the river, shoals of cliff swallows roll and skive. And in the curved inlet, in bright daylight, does come to drink. There is so much life and so much heat.

I am perched on the river's tangled bank, sitting among the mud flies and tiger beetles. It is high noon in high summer. I feel the heat in the belly of this valley. I see the hilly head and its bleached-grass thatch, and smell its sky-blue parched breath. This is the elegant flag of the earth, with its sun-warm and glistening celadons and its flat soil tones of tranquil ocher and cocoa and auburn.

I am watching an ant in the river. Every one of its joints is in motion. It is not sinking, but rather stuck as if in glue. It is being drawn farther away from its mind, the colony. One event of the hill, ant, is going down the drain. It is too far out in the river for me to intervene. All its acrid smell, all that armor, all that labor from sunup to long after dusk, both those eye cases with their thousands of precision facets, all this jam-packed energy appears to be lost. This ant, this lilliputian, is one wee package and yet no scanty event in the valley—full of the purpose and real work of the planet. As you begin to comprehend its significance in the forest, it would first appear that something very important were being lost in this ant's drowning. And yet it is also becoming clear that something important is about to be gained. A body is tossed on the refuse heap, but a possibility is added to a savory tea.

A cricket has lost its foothold, knocked off a weed stem. It has toppled into the river, and it is headed downstream. It swims upstream as if it were being scalded. Muscles contract and the cricket bobs in the current. With its feet underwater, the music it senses might be of continual crash. When it hit the river, this cricket was lost. Just ahead, next to a large stone, there is an explosion about to occur from beneath the surface.

Dead, a bird rolls in the current, spiraling sideways, over and over. It is a fluid dancer. Then it sinks in the current and is lost. In death, it appears fresh, aquatic, broken free from convention. Who knows what turn biology has taken in this striking transplant?

Among the stones of the river's edge, there are the claws of a living crayfish protruding from under the edge of a rock. I glance at my plush hands, warm and supple, and back again to those chiton claws, inelastic and osseous. Both claws and hands are designs that are older than mountains. The dialectic of commonality and difference liquefies and becomes an open stream. Any hope of straight-lined, hard answers is pigtailed. And yet the associations are extraordinary and real.

Perhaps nothing endures but change, but the adage is too small, and the change exquisite and vast. In every twist and turn of the river, there is a very pliant imagery of change and order. I am always coming across a bloated fish, hollowing and bony, full of as much life as it had inside the river as a taker of life. Now its oily, fetid molecules rush at my nose. Decomposition does not neatly summarize in an end. It splays out in a watershed way as rebirth.

The bones of broken plant stems sweep down the river. They wall up on the surface against a dipping branch. There they form a burning screen, soaking wet, yet crumbling to ashes. Like the decomposing fish of the riverbank, one current, a life in grass, is taking another twist. At first it wears the appearance of a clear end because the direction of the ash cannot be predicted. But this ash is as rich as a house of soil, and so there is an ambiguity more than either a clear end or a beginning.

At high noon on the river, in midsummer, the belly of one fish is bulging and satiated. A grasshead swells and droops over the brim of the river. There are thousands of other grassheads taking the same bow. A wild coneflower on the goat-hill vista sticks its face fully into the sun. Across the river a fencerow sinks into a meadow and drowns some distance off under a bank of trees. There is no recession in the valley, which overflows any boundary that tries to fence it. There is seed for another round of bread, and berries rot on the stem and drip into the soil.

There are hundreds of flies drowned and gathered as an

island against a stone in the river. This capacity of all landscapes to throw this precision away is an astounding one,
but perhaps not as astonishing as my capacity to overlook
it. I find the valley to be putting all this precision into another
channel so that nothing is lost and there is movement, in all
events, forward. Just above the night's dead, a fly rubs its
front legs together. Its wings glisten in the light of the morning
star, the sun, like an oily rainbow, as dead flesh prepares to
fly again and keep this star's evolution moving in the activities
of the rivervalley. How can I discard this stream so easily?

In the valley castles are being built out of next to nothing.
They can even be floating on air, looking like dust, but still
containing vast landscapes—infinities of smallness. Precision
is being born out of a few drops of the wisdom of the river.
And from a few drops of rotting fish, larvae transmigrate to
flies, so that even from maggotry, the fantastic. Under every
dreggy stone there is a dramatic shift in the current. Kick up
the dust or raise a small fuss, and I am likely to set off a
miracle, and if I do not set one off, I touch one for sure. In
trillions of cells DNA is a miracle stream that is beyond me
and still me. All over the rivervalley there is this shared map
that looks like a tiny spring far smaller than my unaided eyes
can detect. When I go out into the rivervalley, I am really
swimming back inside the long body of myself, with DNA
being one of its obvious expressions. And by my openness,
perhaps this map is reading itself and laying foundations for
the future.

Ripening river grape gives off ethylene gas. The fruit flies
swarm about these vines overhanging the bank of the river.
The colors of the valley have moved from green to saffron
and amber and plum and russet. There are chameleons on
the back of this chameleon. There is the crab spider huddled
inside a flower, changing from ashen-white to buttery yellow.
Large cumulus cloudbanks build as the warm air gradually

drains out of the midcontinent. The air is crisp. Yellow witch hazel blooms late, and curled leaf boats are beginning to go down the river regularly. The ladybird beetle, bright-orange from the pigment derived from its diet of aphids, sluggishly joins ten thousand others of its species and balls up in the leaf fall. And half the young geese that have survived the summer get ready to join their first migration south over the valley.

The river, then the forest, elbow, then straight, panorama and detail: This back and forth sauntering affords a new angle, like the skimming of a snail that goes upside down now on the clear water film ceiling of the river.

The river appears to end downstream in a tangled and steep impassable shoulder. But when I get around that bend in the river, I know that there will be another straight reach and another bend just a bit farther downstream. Each turn is like a fresh beginning, like a new door. The river's bend stretches light to the point that the river shimmers like a taut band of metal. Everything in this bend is suddenly round, curving and spiraling. Then the straight reach of the river is a straight glistening blade of water. And when I pick up my pace and go from the compact elbow to straight run, the little tendrils and the petals of delicate flowers are lost. In these long reaches, it is the whole view, such as the curve in the riverbank or arching treetops or open stretches of sand, that trips my eyes.

When I come around a bend, I find more river where it had visually appeared to end. My sense of what a river is had led me to expect more river, and so I am not surprised, and feel the wiser for the confirmation. But in the straight reaches of the river, there are small chips of light and sound emanating from everywhere, so much so as to overwhelm my senses, and, accordingly, my self-esteem is whittled back. In my sauntering downstream—straight-running for a time and then elbowing—I encounter the grand complementarity of nature.

Elbow and then straight-running, the river goes on playing this game of give-and-take all the way to the sea.

The transition from detail to panorama abounds in every season in the rivervalley. As I walked one wintry night in the valley, a second snow of the season moved over the hill. It was a dry sand-grain snow. It caused a fizzing and cracking on the dry leaves. The air seemed to pop, too, dry and overcrisp. My breath absorbed instantly into the night. Rainbows encircled the stars. I looked down and sensed meticulous details, such as beads of snow rolling off leaves and fine animal impressions made visible as they filled with these beads. Glancing up into a vast sky, I was aware of an absence of detail. I glanced down and studied, and I glanced up and sensed. Trees sandwiched in between this up-and-down wandering. I drew inside a thick coat and left out a numbing face. I heard the fizzing snow energy and a crunching footstep rhythm. The landscape of nouns melted into wild verbs.

The long-standing cultural view is not merely of things, but of a process of being—things as fluid in time and space. It is perhaps more strongly a feeling than a knowing. In this snowfall or in the midst of the high summer, there is a pulsating and a streaming that *is* being. All events are changes, a rivering. There is a perpetual motion that is not a machine and that does not have to be invented. It is in the activity of everything. Thoreau went out in search of the hound, the dove, and the bay horse. But baying—the activity and not the object—is what allowed Thoreau to travel a good deal in one place, to encounter a cosmos of action in each event.

There are two complementary approaches to perception. One is logical, rational and linear. It is profane. The other is analogue, emotive and horizontal, in which even opposites merge. I spend my life on the logical, and yet there is a rich and creative way of thought that can enhance the rational. It sees order, and yet it also sees chance. It is a means to step beyond "knowledge," which may not really be the world at all, and, in a very real way, blinding prejudice, into the immense complexity in which I am immersed.

By the river I go back and forth. A large shoal of lucite-like needles darts from the river's brim into a deep shadow. By its shadow a cloud is glued to the river. This shadow shape on the river is an alloy cast from light and object. This cool event is made by the same sun that forms a broad sea of white-hot light around this patch. The real form of nature is both, not an either-or. In any landscape the forms of shade remade by the moment are real, and yet they are events that cannot be picked up. Here is a soft complement to the hard. Here is a less focused terrain. Here is a common event, found everywhere, yet largely unseen. This is a beautiful slow dance across the city lawn where I read the Sunday papers, and it is a beautiful slow dance in this freewheeling rivervalley. The whole earth turns, and it reads in the valley on every form and in the changing lawn patterns. I lay down the paper for a time and give up the national news for the cosmic record splayed on the lawn. The shade around objects reveals the way in which there are no blank spaces. The land is whole and deep. Shadows are star patterns. Through these shadows I can see the object begin to crawl out of itself.

Like back and forth sensing, there is always something beginning where something appears to have ended. Water rings go out from a stone in the river. Windblown seeds leave the withering rootstock. The sea is lifted to become a cloud, and a cloud dissolves to rain or snow. Where is the beginning of one thing and the end of another? Each event is not a new thing as much as it is a rebirth. I pick up a stone and I pick up the past. I toss it into the river and I toss it into the future.

At dawn, around a turn in the river, there is a figure in the distant veil of the hill. Its form is ambiguous. It is like a being walking in the fog. Perhaps it is an illusion, and perhaps it is really more internal than external. But it generates in me a visceral presence. It is like the sensation that I experience when I hold a paleolithic spearhead or a meteorite or my grandfather's hand-worn tools or when I hold a fossil. In those implements and artifacts I have a visceral sense of crossing vast interplanetary distances or feeling a heartbeat or shaking a hand that is somehow mine from another time. It is a real sensation more than mystical ecstasy. I am still by the river, listening to the purl of the current and looking into the cold steam simmering out of the river.

As I peer into the veil of the distant hill, it is not apparent whether it is a look back or possibly ahead. Perhaps what I am seeing is a fluid mirror. Perhaps it is a broad bulbous forehead astride a towering figure that is small-toothed, small-jawed, and bald. Perhaps it is my very distant future. Or perhaps it is my very distant past, 150 million years past, a form standing in the mist in a place of very different sound and color, standing near a lush and sluggish river. Perhaps it is a dinosaur, armored, a forty-foot neck, fifty-five-ton mass, something sealed deep inside itself.

I am derived from fish bone and a ganglion nerve chamber that specialized into a brain. It is a corridor for thought, especially useful for time travel, new passage.

Back and forth. The river is planted on its west bank with dense willow and cottonwood, and on the east bank by chains of grape. In high summer the grape hill is dry and dusted, and the willow copse warm and moist. The river joins opposite to opposite. Seen through the willows, the sun is broken and diffused, as if it were passing through a willow screen. Seen through the grapes, the sun is broken into white-hot spears.

To the south and low on the horizon, contrails cut the red heart, Mars. The moon rises on the horizon, distorted by the atmosphere and swollen to fill the sky. It milks this valley and my far-off cottage in the city in one simultaneous light.

The Big Dipper has dropped down from its summer perch, and it is low on the northern horizon. At the end of winter's first day the crows coalesce again into rookeries of wild brilliance. Snow has yet to curve the red-slatted fence, but the night is coming two minutes earlier and staying two minutes later. The leaf buds are waxy, and the frogs have given in to deep slumber.

What promise lies around all bends, all cross-cutting, all straying? What is it that all exploration—all passage—ultimately pursues? There are a thousand meager reasons for going out and penetrating deep inside the landscape—curiosity, fame and fortune, desperation, solace. Yes, it is true that when we go out we strive to change, but it is to become somehow more sound. My means may appear revolutionary, uptuning an old vision, but the end is evolutionary—to re-vision and not cast out, to remain within a continuation of being.

Permanence is not really an endpoint. The highest mountain range is one ancestral seafloor, and in no time at all mountains are fleeting rainbows. In no time at all leviathans shrink to lice; yes, the ones singing every morning just outside the window. In no time at all your and my eyes have journeyed from the side of our heads nearly to collide in the middle of the face. In a very short time seeds become towers and crash. Decay becomes recomposition. And in next to no time at all tasters flower into the tasted. With all this change, what is it that is permanent? While landforms dissolve and reform, and our bodies meld to new form or perhaps fly to extinction, there is always that strong ambience of continuation, of passage. Across epochs and in no time at all, what is never absent is the passage, the river of change, sometimes looking like rain or plants or animals and sometimes looking like stone and empty space.

One obvious name for that which is always present, and yet always fluid and moving forward, is rivering. There is, of course, that remembered river of childhood and the new ones of my city. There are slightly less apparent rainstreams, and the veins of leaves and hands, and the shadows of trees. There is the stream of the seasons and of the wind and the flight of stars and the movement of the landform in erosion and drift. And I am reminded always, with each breath that I take, of my rivering.

I come down the hill to the physical river of this valley. I stand by this coastbound train, and I imaginatively step on board, and I am swept around the bend. And when I look back upstream, I am entering as rain that has come all the way from the sea. All events fly forward, but their complete essence also involves a process of returns, of come-arounds. All events are a circle of continuation and permanence. I stand by the physical river as a watcher, and when I look deeply, I find that I have always been on this journey. I sweep out in my physical respiration and in my perception, and my water body stream is a spinning water planet woven to its star stream. I find that I am not simply inside a landscape, but inside my long self, which is really to say, inside my permanence. The physical river is before my hands, but I am deep inside the river eternal, and it is real, not imaginary.

Without traversing rare ice fields or the harsh distance of interplanetary space, the profound journey to the eternal is in every local place. Without taking a step, I am on a journey that has long been under way. By taking conscious passage of that journey, I am home, homing—less a place and more a process. Then why so long in coming? I have been approaching the landscape as if it were an adversary. By always moving outward, I can sidestep the experience of being inside the landscape. It is easier to go out than to face the less familiar reaches of self. Home is both veiled by my familiarity and by its challenge to my basic strategy of separation and dominion.

If this view comes hard and the most that I can encounter

seems to resemble a veil or a ghost image, that view will likely be enough. The river looks the way a real ghost town ought to look: doorless, windowless, unweeded. And the ghostly image that seems to be there moves like one, in overturning power and tickling purls, incessantly remaking its spilling form. For a long time I had not believed that I could digest a river in any other way than by taking it apart. Now a fish's sword-flash escape from the fishhook can flower to more than a lost meal and become perhaps as much as the fodder of wakefulness. I can catch myself, consuming my listlessness and my narrowness. The perception of ghosts may just be the first step to break the iron wall of the familiar. Perhaps the ghost will be my final bastion against seeing that way in which self is a river.

By listening more than analyzing, I encounter a broad prairie river that is muddied and laced with willow and foxes and pheasants. I fly down the hill to the grassy, tangled bank of the river, with eyes blind to the eternal. A muscle against the muddied, greasy slant or my ear or perhaps my hand in the current becomes the intermediary. I have believed my eyes and my hands to be more than a river, to be, in fact, the river's divine leap forward. But in the force of one small river against my hand, I know that I have been haughty and limited. The river has met the waters of all rivers—the Ubangi and the Marne, for example, or the Ucayali, that Peruvian river that joins the Marañón to form the grand Amazon. And by the measure of time, my lifespan, in comparison to that of a river, is by far more brief than the life cycle of a mayfly.

I am drawn with a sense of homing to the river, like steel filings to their charged bar. I can envision my own private moments as a river. My life can be sensed as a rush from the river top—egocentric, then detailed—to getting lost at sea, and diffusing to the point of no individual identity. And yet, even then, I continue by contributions and by my abuses. The body of time spent reading this book is a gun flash; and a lifespan a get-rich-quick scheme. Surprisingly, it remains much more difficult to imagine how my everyday, which is largely outside this valley, can be a river. At one moment culture seems to be too far ahead, and at another too ignorant and too inconsistent and too self-absorbed to match the wisdom of the rivervalley.

The harsh metabolism of the city and my own high-speed metabolism have never left the grand river of natural process that I observe in seasonal cycles and stellar evolution. My city life, my everyday, is perhaps more like a wild young river, infantile, having been in existence for only one-fortieth-millionth of the time since the formation of the planet. Its "artificiality" is perhaps wildness itself, a miracle of adaptation and perhaps an opportunity for long-term habitation, all of which may make the city stand at the vanguard of nature. There is a quality of cultural continuation, unbroken from the paleolithic through the post-industrial. Like a river, its essence is not fundamentally reducible to a set of characteristics. Still, there is always that opening smile, that ecstasy of dance, and cycles of transition. This cultural stream slow-grinds like a continent, changing more as stone changes, evolutionary rather than revolutionary. It can be listened to like a river, but it cannot be accelerated. It is too big, worthy only of my admiration. The great promise of the physical river for my everyday is the capture of its rhythm. It can be nurtured as a grace in the everyday, to catch the rhythm of the underlying current of culture.

I sit down by the river, staking out a perceptual weir to snag that which is coming downstream and forward. By my returns to this river, I no longer stay a bench dog. I am stolen. That which sweeps me off my feet is a landscape of vagaries—erratic and unpredictable and vitaceous—but also a rich brocade of silk and gold that warms the measly part of my beliefs and vitalizes the twopenny part of my values.

On the river's tangled bank all plants are Turks' crowns,

and the summer hill a belvedere, and the sweet night grass a sahib's charpoy. And inside the river, in this kingly sized reality, there are a hundred million miracles. One miracle: as squiggly hatchlings or mature, slow-rolling brocaded logs, carp; they blow away from the edge in a whipping puff. They are like a life inside the floor of a temple. Their books are there with them, unwritten, not needing to be written to speak openly to me. In their simple endurance, their aesthetic and philosophy are on parade. They slide solo and in choirs unfolding in the current, moving like my deepest quiet breath. The way that they live under ice and in the flood and through those molasses days, it is the life of a rich dream. They are like the civilizations of my dreams, choosing to live simply to live fully, and operating with usufruct, using events without altering or damaging their substance, which is, fundamentally, permanence.

There are a billion more obvious stories in any river. These tales look like objects and sounds. I let the river itself turn the pages as I sit on the tangled bank. I can leaf through grand novels or through quick-glance sideshows. Even the sideshows will still be more than enough, and they can tell stories of culture as well. There is, for example, a spongy shoe—a civilization of fungi—and a light bulb looking like a bobbing dead eye, a 1924 quarter and a wedding band, and a stainless-steel single-edged razor. There is a Christmas tree story and a bedspring that remembers everything about a marriage, a broken clock still deep inside time, and a telephone and a telephone pole. There is a hot pistol in the mud and a postcard from Maui. There is a bee and a muffler. There is a fragment of wallpaper consisting of delicate rose patterns and it is flying forward. There is a blackened hide and rotting eyes, a cigarette butt and a spent bullet head. There is a bubble leaping and a cherry Lifesaver wrapper and a broken-tipped dagger. Is that a master or a fool watching, reflected in the water? And perhaps a mile away on a dead-end country road, is that a horn's honk falling into the water?

There is a time for short stories and novellas, and there are dazzling ones inside any river. Perhaps they are the easiest way inside this landscape. By going to the obscure and the tiny, I am again shown that in the small, vastness is revealed. For like the small sideshows in a river, a vast planet is a star's anecdote. In the power of the river, all the speckles deepen the wizardry of that tea—all those throwaways and grains of sand. That which keeps us from finally coming home is predominantly our capacity to dismiss the landscape. The strongest veil is not as much distance as the subjectivity that familiarity breeds. I am more than willing to dismiss just about everything as insignificant. And yet each sand grain is moved by stars, and is, in fact, a star's activity. From the smallest detail to the sun overhead to this river, that was a coal swamp and a sea and a glacier and now a prairie river, all are passage, forward-reaching and coming around again. Nothing is important and nothing is unimportant.

Going Inside

First, one cautious step and then two, I ease down the riverbank, unbalanced both by the slant of the brim and by the flow into which my legs are spooning. I step out farther from the muddied bank until I am chest-deep and walking on flowing sand, and being pushed gradually downstream. One more half-step toward the center and I am swept off my feet. Suddenly I weigh next to nothing and I am moving forward without taking a step. The blazing hot summer rim cools out behind me. My left shoulder rolls forward and my right foot backpedals to counter and balance. There is the slight, anxious tension of expectation and readiness on my lips. And for my eyes floating on the surface like new lost seeds, the view is wide-open and refreshing.

I drift somewhat tenuously, with my torso and appendages protruding through the top of the long tea-brown body of the river. I hold my breath and sink inside. The landscape of the interior has no air, and yet my perception seems to grab onto only animation and spirit. As I glance back up through my squinting eyes, light rains down like warm arms, and sparkles with a living diamond dust—drifting plankton. Current charges this channel with force, and all things are spiraling, turning like prayer wheels. Only a few large stones in the sand appear still. There is no furniture. This is, for me, a body more than a room. There is a pulse. It is both the fresh wave of the present moment and the ancestral carried forward.

The dead fly downstream around me, going forward perhaps more than descending into an underworld. This river way is less like a River Hades and more like the River Ganges, the door of heaven. My river has rained out of the heavens. And inside this river, the mewl of the newborn dances out of the dead. I see the drowned vegetation now transmigrated to spawn, being carried by the thousands under the tail of a crayfish. Their mewl is a song that says the dead are in the long body of the living. It is a litany that says, over and over again, that existence is an arc of ascent and turning back down, and it is more. Being is, in a simple way, a circle. It is, in a more complex and fuller way, a spiral, not a repetition, in which the energy of the dead flies forward to live in new color and sound. And the evolutionary change that occurs in this forward spiraling of existence is a stream of slow creation.

When all else fails, there is this moving landscape, the physical river, where I can go in increments of an hour or a day, a month or a lifetime. I go there and my bearings and my headings are regained. While others are busy at the stream, I do next to nothing. I try to behold and listen to that which is already being done.

I am drawn to the river for more than its honeyed sound and color. I can find that in a hundred thousand places. It is there, for example, in the sheer cliffs of Lhasa and in the flower boats of Xochimilco and in Mount Palomar's lens. I am drawn to the river by its undisguised fluidity. Inside the river's streaming, I rise out of my gravity. Occasionally I may literally fly inside the river. More commonly, while grounded on the riverbank, my perception takes flight in the fluid imagery of the river. The young and old roll and mix, long-running and then slowing down to a crawl, adding sugar to the tea. By the river, I am calmed. Against my moments of routine and crisis, I feel as though I had been given an elixir of recovery.

I have gone inside the rivervalley's clear and ashen days and some of the incisive whorls of its seasons, which is to say, inside the opportunity for respite that the river offers up easily to me. Respite from the thirst of my everyday, from all those moments of robbery by petty demands and solid necessities is enough to ask of any landscape. In fact, a demand for respite may be simply another "use" of the landscape rather than simply my admiration. And yet, by my returns to the river, I have been taken further than respite. I have been lifted out of the gravity of my preconceptions about self and place, and I have been swallowed whole by the river. It has been as if the river had recalled me from a self-imposed exile. And that which I now go inside is a larger terrain, out of which I can never step. I am inside the landscape of a physical river, but also inside its larger name, river eternal. And I can no longer use it because it is my larger body, and I finally experience, perhaps more than I understand, Thoreau's admonition that "this curious world which we inhabit is more wonderful than it is convenient; more beautiful than it is useful; it is more to be admired than it is to be used."

When I go inside the river, whether it be the local river or the larger name of river, nature, is it somehow out of one world, culture, and into another? Once inside the river, the tangled bank does not seem to be a sharp boundary. The river still expresses the roll of the planet, the seasons, and it is still deep inside the roll of a star in its galaxy. Drifting in the river, moving at perhaps two feet per second, am I really leaving anything behind? For myself, the most realistic distinction between the hill and the river, and even the city, is the degree of clarity of the bell sound of nature. In the city the houses and street are a rackety dream, while the most unnoted event of the river resounds clearly and deeply and awake. By its subtle trill, for example, a needle-fine water grass, spiraling in the current, slakes my thirst for wakefulness, far more than a gaudy billboard or the fashion of the moving mob.

By my returns to the rivervalley, by going inside the powerful clarity of the river, my vision of the landscape has progressively widened to the point of coming inside my own identity, and I inside it. And the river has broadened to the point that I can feel its flood washing through the aspects of my everyday. Now, whether I move softly downstream by or in the river or wander the most hard-edged roadway that the city has devised, I am deep inside a common ground. The rock outcroppings of the rivervalley and the rolling shoulders of my capital city contain casts of geological change—seas and soils and glacial cobble. And more than remnants of another world, they represent an unbroken, common process of continuity, which is also inside the present. The life of the goat-hill prairie remnant of the rivervalley sustains, not merely as a vestige of the past, but as vital and up-to-date as the life of the city, and both sustain as expressions of a vast global and cosmic wilderness. And the essence of both contrasting landscapes is a river, eternally streaming.

I do not have to go to the river to realize a sense of place expanded from my immediate region to a cosmos. Without taking a step, I can sense a relationship with more than culture. And yet, by my returns to the river, I am cast into a view of nature that is far more compelling than a relationship. By my returns, I experience a nature that is more than surroundings and more than a physical resource. It is a landscape of process that is not secondary from my everyday, not separate from culture, but vitally active in the metabolism of my city life, especially visible in the urban variety and renewal.

River eternal is not simply another name for nature, but rather a name for the *experience* of inseparability from the landscape. It is an expansion of identity, in which a person becomes a place and is expressed in the events of place, regardless of how minuscule or how distant the events may appear to be. There is a rivering, a streaming, which rains simultaneously through all the events of a landscape and through the vast wilderness of oneself. River eternal washes broadly, like the river of the Everglades, quenching all at once. Within and without, there are nature and wilderness, yes, but it is also homeland, the larger body of self.

I ease down the tangled bank of the river, and I fly down into my brown-green body. And without going this far, without even going to the physical river itself, I have been taking a grand rivery flight all the while. It is river eternal, the quintrillion channels of which I am but one wrinkle on a wave. Fully in touch with the river eternal, fluid and changing, the physical river of any local place or the rivulets or the veins across the back of a leaf or my hand are the clearest shortcut that I have found. I listen to these musics and then I go back inside my everyday. I listen for the living metabolism of my city, and I begin to experience the sense of respite that was easily opened by the river. And more than this, I begin to find a city inside the present one that offers a sense of homeland more than exile. It is that rich, enduring stream of cultural continuation that derives its currents directly from the earth.

There are a thousand flights to be taken inside that grand illuminator of river eternal, the physical river. There is, for example, that imaginary flight, lasting less than a minute, over the lip of the dam. It starts smoothly, steady and straightlined, rather than in long curves and the quick midstream whorls of the unchanneled free stream. Then it is out suddenly, over an edge, into the air, and I leap and fall into the wash, and arch up in pressure toward the sky and back over— all entrance and no exit. Water flies in from downstream to fill in the hole made by water exploding downstream after the plunge over the dam's lip. And there is also the slow geological flight, in which shell fragments and cottonwood seed fluff rise out of the elements to form whole mollusks and towering riverbank trees, and then fall back to elements again, and further still, out across starfields to become stars again and stellar death beyond that. And sitting as still as I can, I ride this planet in its flight, a particle in a starry current, with mountains shrinking at this vast magnitude to far less than the height of wrinkles in crumpled paper. And I go with all the seeds and bones and feathers and bottles that are flying down the river—trash, yes, altar pieces, yes, and vehicles in flight. The most solid-appearing object is energy in the vast spacetime stream, as liquid as rain. All is a river.

I am watching a wood thrush, listening with its feet to the living stories of the leaf fall. I can hear these stories, lost to my ears, through the actions—the commentaries—of the thrush, in the turns it takes here and then there. It is spring, and a small pool of chorus frogs sings like perhaps twenty thumbs stroking fine, long-needled combs. Their song sounds like a secret, and yet it is a wide-open chant. Perhaps I am the event that is wild and remote, and this obscure valley calm and civil. Perhaps I have been looking outward for too long, not listening to my heart, not listening to the river rolling like a calming influence under the traffic.

I begin to watch myself. In my movements I begin to see a landscape swimming through a forest. I stop among the tall fronds of the ostrich fern, hip-high. There are no words. I am speaking in semaphores, gestures, in slow lines of movement and acrobatic quick-turns and in touch. It is a mother tongue, unknown to me, even though I am speaking it. In my respiration there is my mother's lullaby, and through my hands my great-grandfather moves, and in my moist swallow the primordial worm. I am no longer immersed in a garden. I have become a current. I touch a tree and a double entendre occurs, a murmur of primary patterns, a swirl of gravity and radiant energy, inside the riverine forest.

When I cast my hand into the river, I am coming to understand that it is a hand thrust into my pocket. Something is moving inside me. It is the nymph scurrying under the rock. It is the microscopic diatom on the filmy surface of the river. It is a shoal of translucent young fish. These events are aspects of myself. And a river has me for two of its heavy-headed hands. And when I take my hand from my pocket, there is a wave. It looks like next to nothing, like a small gesture, but it is one soft roll of the galaxy. This is not poetic imagery. This is the unabashed wonder of reality, the reality that runs the planet while all the books are being transcribed from muses and edited for the hungry market of dreamers.

It is not alien, this river. It is my tear and my emptying vein and my reoxygenated artery and my voice on the loose. And it can tickle me and be the cold justice of death, no, more the joy of death. It is to where all death flies if I give it time to stream there. It is that to which death aspires to become—to be the living edge of the flame, not some burnt ash, as I have wanted to believe. Raveling, a river is both one of my new short stories at every moment and that final hymn of mine at dusk.

My life is never so private as to be removed from the river, never as separate as I would like to presume. My most private moment, and also yours, is always a river's current. One moment of my life is a splash; one lifespan a rich seed. And no matter how far I will rise out of the earth in my distant future, I never fly out of this stream. And my everyday life, no matter how refined at one time or how low and blue at another, remains buoyed up by this stream. My life contains the same problems of the river, the bridging and damming and channeling. And it also contains the same wild, brilliant reality—the continual creative twists and turns and the stillness and the unexpected falls and all the wide-open sunlight.

As the river of this valley carves a vast watershed, so I am carved by the river eternal. As the local river is nothing but sea and air and fired soil, so I am also contrived of more than my appearances. As the river weaves back to the sea, inseparable from it or from the hills and the atmosphere, so I weave back inseparably deep inside river eternal. No matter how freestreaming I feel, I also know that I am connected to a river eternal, not in a relationship with it but as its nuance, as a subtlety of its grand fluidity. Instead of the image of a river as a place, I increasingly wake to the *action* of the river, and of any landscape including myself, as its essence. And that action does not stop at the bank, but sweeps up over the bank and inside everything.

A wave in the stream can be seen as a mechanical process and compared with forms such as, for example, an uncoiling seed. In a circumscribed sense, a wave does share this symmetry. But when I experience a wave, and then another and another after that, there is no bone or coil or soft heart inside. It is the water and the wind, but its truer name is energy and it's not even energy that is dissipating in one direction. There are waves all over the surface of any one wave and waves within those waves. And what does all this diversity of the surface of one wave suggest about what is occurring inside that small rolling hill of water? I look closely at this wave, and it is a madding texture. Science is here, but that science is such a pure, rarefied game in comparison to that which my eyes scan. Wave science becomes a loose dance at best.

The complexity and looseness of a wave is a quick glance at the way that all landscapes, including myself, work. In geological time, whole galaxies birth and age like one of these small wrinkles that last for a split second in the crest of a wave. The smallest local wave is inseparably bound with the cosmic. I know that if one small wave in one local stream would stand still and I could pluck one thin wrinkle on that wave, it would be a cord that winds back through that long body of the river and back into the draining hill and to the emptying cloud and back into a burning star.

My account of the details of one wave is not intended to be an anecdote to the river's tale. Perched on the river at dusk, watching the wind in the waters, I was astonished by the complex nature of a wave that was revealed in the backlight. It seemed to be saying that the essence of form was activity perhaps as much, if not more, than structure. The essence of all landscapes, including myself, was a streaming, a verbal form that can be seen as a noun, but not fully comprehended as a noun. Eventually, if we are to sustain as a species, we are to be overcome as a measure of our success, as we overcame our predecessors to become species, *sapiens*. We are to be overcome, which is fundamentally to recognize that it is not our nature to remain static. Our essence is not form as much as it is activity, and its range is far broader than we have allowed ourselves to imagine. I am a wave, rising from a dissolving wave and again dissolving and carrying forward in a wave that, by its inseparability, is also my streaming body.

Before I stepped inside the river, I presumed that I would be taken into the presence of other nations. But when I stepped down into the current, I slipped inside a view that I never intended to encounter. In that slip my dominion was taken, my eyes were washed, and I began to see that I have forever been inside the river eternal. From that point on, all I could see were rivers in the grass, in the wind, and even in the metabolism of a city. Still, this view seems, at times, too intoxicating to be real. But I am coming to understand that my questions about this view are about the depth of poverty of my perception rather than a question of the validity of this view. I have not been capable of illustrating just what a wondrous landscape I not only inhabit, but express.

Respiration and digestion flow in waves through my body, as if my body were a river. I carry the map of some obscure river on the backs of my hands. And I am made from countless other maps that form every bend and straight run of my body. Perhaps every detail of every river bend and every fluttering rapids can be found in this internal landscape of rivers. The similarity between a river and myself draws me unwittingly at first, but strongly. And when I go to a river and I swim with my senses inside its long heartstring, I seem to complete a circuit. And the real work of the world—the rivering—channels through both the river and myself, and I am struck awake by the clarity and power of a river into a consciousness of this rivering. And when I swim out of the rivervalley and go back inside my everyday, I can no longer forget about the river. I catch it in my respiration and see a river in a leaf that is fluttering in the backlight and in the outline of the tree against the darkening sky in early evening, that seems to be a river beginning in narrow rivertops in the sky and thickening and draining into the earth. I do nothing but behold rivers.

No matter how far out from myself I sojourn, I find that I am still deep inside myself. By coming over the moon and viewing earthrise, what changed, for me, was not the depth of place as much as the depth of self. In that swirling river of atmosphere and ocean and landform, there was only connection. Instead of being lost, I was home. I cannot truly convey to you the joy that this revisioning of place generates in me. I listen and do nothing and I quench my thirst for more. The landscape spills over and rains into my senses.

What value the smallest event or the most everyday event? The wind carries the risen sea. It topples off my roof in the city and sweeps down the hill in this valley and rushes across the open surface of the river. The ripples in the surface are the visible sweep of the wind through the entire watershed, the washing floor, the rooting ground. And when I go inside of even the smallest event of any landscape, this grand view floods my senses.

Before my returns to the river, the valley had been a small

world of my margin, lost for the highway, on the low side of the terrain. It had seemed to be tucked away like a brown cellar. It had looked like next to nothing, no-faced and meager. But it is a landscape that can take a city apart. It has carried away mountains. And although it has appeared buried, it is nothing but sunlight, with a profusion of flowers and deep greenstone ravines. There is the rich variety of the wind, rippling the river and knifing through the willows and fluttering in the high cottonwoods. There are the longstem grasses and herbals of the tangled bank, not taking as much as readying to give themselves away. In all this wonder, the bell at my neck—my throat—is silent and listening like my eyes and ears to this brown tea river singing. And if this neck bell does anything, it takes in this river's resonance and carries it back inside my everyday to soften and deepen my words.

There are turns like this river's edge everywhere on the earth. All the millions of leaves of these anywheres wave like truce flags. They give themselves away, like water flowing downstream rather than hoarding itself in one place, remaining fitted and unending and inseparable. All this wild wave in every local landscape is not stoic but open, and it is boiling energy. I can turn over a dead leaf or a stone and discover an aspect of my life there that looks like the conjuring of a mad god of magic. And out in the full sun, Dionysus, the white-light intensity, is there as the brilliance of noon.

The river says "I" and it says "sea" and it says "sun" when I open my senses to it. I rarely have to do anything except behold and listen. Sometimes I take away a sense, such as my vision, to unplug blockages in my other senses. There are too many moments when my senses see the river as a line contained. But when I open to the activity of the river, for all its apparent narrowness, it is nothing but room. I cannot fence it. The river is the visible tip-top of an iceberg. It is the visible head of the sea on land, and more, a cosmos's energy being expressed. The only small thing in this landscape is the anchorite—my meager perception. But by my returns, my senses flower into awe and enchantment, and I swell to the magnitude of a star. I can be home, back inside the long body of myself.

The bones of the earth seem to wash down and settle inside the river. But when I sit and behold, they fly awake and overflow with possibility. Soaking wet, these bones adorn, not desecrate, the heart of the river. The coldest bone is a burning fire in the cosmos. It is there in the river or in the forest like a venerated artifact, still very near the beginning of this young planet's life. This bone is more of a blueprint for the future than it is a finished product. These are not the bones of the dead. They are more like a living transubstantiation, like a caterpillar transforming to a butterfly. I come back to them in a day, and they are half-eaten, and by the next season they have flown off like a bird in migration. And like these bones, I am not as old as I would like to believe, not that finished.

The form of the physical river of this prairie watershed: I see its thousands of tributaries as arms reaching out, with a hundred thousand fingers each, and the drops of the hill like trillions of fingertips. The river is touching everything everywhere. Like the sun, that does not stop at its photosphere, the river is not bounded by its bank. These are simply the points where my eyes are likely to give out. Like the radiant energy of the sun, the fingertips of the river dry and then reform moment by moment. They shift and dance under my feet, like the more visible trails of ants and the territories of birds. The river is interwoven with its watershed with a delicacy that is like the dance between all wasps and their flowers.

The form of the main channel: It is the long heart of a lost glacier, more fluid in the summer when the tea is warm. This channel is that which I commonly name river, and it is that which distinguishes the river from other water bodies. For its indigenous plants and animals, the main channel is perhaps like life in a moving vehicle—a gypsy's caravan. Its drift life moves at perhaps an average of two miles per hour, forty-eight miles per day, slowly down to the lethal salt. This channel expresses the art of tea and waves, perhaps appreciated most by the clam's blind shining. And all the variations of its flight are, for me, a trigger.

A river is as many things as there are perceptions. Its value to me? When I grab hold of its tail, it is no longer simply water "out there." The river is unthinking virtuosity that is beyond my capacity to preserve or benignly steward. It is a tea of wakefulness, especially to that way in which any event is so much more than its appearances.

All my thoughtfulness seems to make me so much more than any river. It makes me want to stay inside, behind my eyes, and play. But the river "out there" in the lowlands is more a construct than a reality. For far too long I have sensed the river to be a dead end, a drain, or a watery road at best. I have been asleep, perhaps dangerously asleep.

Yes, fitted like a hand in a surgical glove; yes, I can imagine a mind behind my two bright eyes. And yet, by my returns to the river, I am beginning to comprehend that to stay really inside the mind is to go out. And when I wake and go outside, I do not leave this mind behind. Looking out from my two optic mountain caves high above the soil, it is my face beyond me that I see in the landscape. This terrain, perhaps a neighborhood or a national forest with two lakes and a river, is the undiscovered terrain of myself. It is more than another nation. When I had first thought that I had heard the wake-up call, I saw this beautiful other, nature, as separate from myself, culture. And yet this is only where the journey begins. By my returns to the river, I was literally brought to my senses. To imagine magic is to be readied, but to be present and open is to be fully alive to wonder.

When I wake up and go out, looking out from a mountain

peak or a hilltop or a riverbank or the window of my city cottage, it is my face beyond me that I see. One moment, from my cottage, both my bright windows look out on this face, and it is a church on a city hill. And when those windows pivot eastward, the face that I see may wear the body of a man and a dog on a slow run down the avenue. And when I really catch onto this face, the church and the man and the dog and their hills are just different whorls in one diverse river of nature.

There is that something deep inside, behind my eyes, that is larger than my head. It is the streaming of river eternal, and it is countless times ahead of the way that I am thinking. My thinking is not something that is really my own, or even something contrived in a thousand centuries. It is an accomplishment of the grasses far more than it is mine. Its most complicated word may look as if it could only be mine and not an eagle's or a river's, and yet, it is only a very new form of wind in the grasses. While I may follow the vote on Capitol Hill as if it were the main determinant in my life, the ebb and flow of trillions of cell assemblages within my body are attempting to synchronize with the ebb and flow of the landscape.

When I go outside I go inside, and when I go inside I am stepping out. Because of the way that my life appears to be one turnstile, I have not wished to enter this gate. It appears to be a step backward. It is, in fact, a joyous flight when I take it. At first it was a wrenching shock to my sensibility, but it quickly flowered into enchantment. I encounter not simply a curious quirk to existence but a deep resource to draw upon to sustain for the long run. And with this view culture does not melt away. This view is cultural advancement, essentially a measure of my capacity to step beyond the veil of my beliefs.

Fine handwoven rugs that may be the epitome of a cultural tradition can be found in greater refinement in the texture of the leaves of the rivervalley. And I find other textures to adapt into rugs. I find the storm pattern and the still-pool mirror. And that face looking back at me in the pool's reflection is the river. Why is this view so hard in coming? Well, my eyes seem to be saying that the real life of the earth is broken up. And so, to live, I am always trying to pick up the pieces. I am always trying to plan some master scheme or rearrange this or add to or subtract from that, because nothing seems complete. The events of the world seem separate and in disarray and in need of invention. And I have come to believe that the capacity of culture to step beyond nature and induce order is the reason we have flowered to people the earth.

I know better than this. By my returns to the river, I have developed the strength to pull free of this illusion. On cultural desires, my eyes have been hanging. Now, in my eyes, I find the river's view. No matter how narrow or how opaque or grand or commonplace the window, the view is upon nothing but the infinite. A train passing the window is the whole train's action, whether or not the whole train is seen. And the river and myself are the sweep of the grasses and of the sea and the stars. Now, wherever I begin, in the still pool mirror or in the rapids, there is a current running right through the middle of both that goes back to the main body, that buoys up and makes the diversity of the whole possible. I can go beyond the limits of my eyes, and I can see this current in the activity in the window. I find that I am really not ahead of the planet. The earth has given culture a quick start with the stored capital of the grasses. But now I will likely need to wake up to more than appearances to sustain for the long run. In this view that which I abuse and that which I ignore is also the larger body of self. And with this view these eyes of mine are not lost. Nothing but deep pools they become, these eyes of mine.

Why is it that I live? Why am I here? Why do I bother to wake up in the morning? I have a keyhole view at best.

If I can be honest, I answer by cutting up and discarding all the old questions, and I know this much: My keyhole view is enough. A panorama view of the cosmos likely would not change a thing. It would not be enough, because I still miss the obvious in my too-focused view. The swish of straw on the patio, the light around plants, and the blue sky and the gemstone in the eye: I am willing to keep overlooking all the obvious answers. Perhaps far more than those elemental questions, these activities are the living questions, and they hold that which is life to me. And this awareness remains beautiful always, because what is beautiful deepens and even changes. With time, there is a beauty in graying and in children, becoming us beyond us. This awareness of these living questions—these infinities and unanswerable contexts—is perhaps that which I can do best. I do not run the earth. I cannot even help in the things that really matter. The day gets here by itself. The weather suddenly takes a new turn, and my whole life could change or end around the next corner. I am not that different from the sparrow chattering in the trellis.

My changing view—my sense of rivering—is a wake-up call more than a better method of control. I need to remember who is catching whom. The old man on the bridge in the city and the fish grabbing the hook, the swallow skimming the river's matt for its delectable gnat, and myself—all are certain that they are the one, but we all end up going down the gullet. Who is this fisher that is catching us all up? Well, that which is catching us all up never crumbles. It underlies East and West. It is inside stillness and the noisy rush.

I am in for quite a surprise when I try to put a name tag on this grand fisher, this event that appears to be above us all, that event that I keep looking for way "out there." This grand fisher is far beyond such a duality of "I" and "thou." It will not cut up that easily into an answer of parts. The dragonfly after its boneless gnat, the sparrow after its moth, that old man and his hooked fish, all these fisher boys and girls, all these mouths, are the fibers of this fisher. It is perhaps more of a process, a fishing—a rivering, a streaming—than a focal point entity. And if its form can be known, its fundamental nature remains an opening question, which keeps me in search and awake.

While it is physically harder to journey to a mountaintop or to the high Arctic or to the ocean's abyss, it may be psychologically far easier than facing the way in which any local place is *we,* both self and river eternal. I lean too hard on my eyes. They are not organs designed to show me the earth. I have no organ to sense the whole earth, or even the whole rivervalley, or even one thin strip of the tangled riverbank. With my eyes, I see enough to eat, and I can see nothing but miracles of color. As a hawk's eyes lean toward a range of yellows to discriminate between objects, my eyes have their fixed slants that limit sensory reception. They cannot see the way that the concrete streets and heavy houses float precariously over an abyss. My eyes cannot see that what each hair follicle probes is a freewheeling wilderness. In the midst of the city, wilderness is unseen, not absent. And the landscape that I see is more the one to which I aspire than the landscape that exists.

Over and over again, I have been talking about one thing, not the physical body of a river as much as its activity or streaming. It is, for me, a profound view because of the way that it expands to a grand streaming, river eternal. Rivering seems to be a meager concept to tag onto such a dazzling force. And yet the term is not so much a descriptor as it is a trigger. It is never as much the bony concept as much as endless nuances.

A river has superlative power to override my limitations. It can turn a vast city into a dot on a line. And the way that it looks like a line makes a river into a literal guidepath. It is a pathway with infinities of smallness too deep to subsume. A lifetime could be spent on just one bend in the river, asking questions, opening and listening. And this way that the answers appear to end in nothing but questions is an opportunity more than a problem. My eyes learn to soften, and I begin to allow events to open my sensibility and to fly inside my identity.

By coming to the remembered river or by the local river by which I find myself, I am taken beyond the way that language and thought are typically used. It is the unthinking river that can appear to be the rational one. My too-rational view seems nearly irrational in its loss of contact. But in the events of the rivervalley, there is no decision that does not work or that is a delusion. The nonmovement of the fishing crane and the translucency of a shoal of small fish—the little yet endless nuances—begin to move inside me, and their flight inside my awareness is my strong answer. My thinking becomes more fluent—a dance of awareness more than a mechanic's tool chest. And the images that pirouette and spin out from this dance lift me into that power and delicacy of a river that is solidly present but unspoken. The river may still resemble a string of water and be faceless, and yet still express a thousand gestures at once and not really be either dead or alive, but beyond such a question. My words begin not to have to choose, and can be simply present. It is then that a word stands to be alive and kicking, and I can feed on its edges and be thrown by its spin into the real answers that keep me awake by the strong force of their shout.

The embracing river view is a gift. It allows me to care for others, for that we-ness, for that long part of myself. No difference of nationality, race, creed, conviction, age or sex, or species is large enough to overcome this view. At your river running south or even north, or yours falling off a mountain, or by yours drying in the desert, or by yours and mine chained and discolored in the city, without taking a step, we share a common ground. Within all stone as well as these waters, a river eternal runs. And it washes through the middle of us, and I find my fluidity, which is that way that I am more a current or a color in a current. I am one quick-twist in a channel of this grand river that has risen through time from the grasses.

It is one thing to live inside cooperation and fittedness, and another to realize it. To live inside the forward reaching of the earth may be all the difference, because of the way that it enchants me and the way that it charges me awake. By my returns to the river's fluidity, all that I encounter is transition. It allows me to look beyond appearances and to follow my activity as a current back into the events of the landscape. And by this connection the compassion that I begin to feel for the landscape is a response to my own self-interest. The smallest act of compassion resounds deeply, embracing not only an immediately deeper quality to life but also becoming a gift to my future, enhancing that stream of my species to come.

I have wandered by one very thin sea, a prairie river. I have found a watershed way. Looking into a warm leaf, I see rivers. Every event is there. And I have stumbled onto a star's journey to a seed. A star—sun—and a seed have formed a bridge. Sauntering across it, I have found that I belong. Sound and color have formed a door to the earth at every turn. In the music and pattern of the rivervalley, I find that every event, including myself, is a whorl in the stream, a fragrance in its tea. For all the denegration of nature, here is the beloved homeland still, in all its greeny prime still.

All is a river—time, self, place—flowing moment by moment, day after liquid day. I am inside the present, a wavecrest in river eternal, rising from the past and cascading into the future. There is still the magnitude of wilderness, a great diversity of endurance threatened, a multiplicity of utterance, and an opportunity for responsive action. I hear my mother tongue in the purl of the river, in my respiration, in the wind, and in the patient silence. It is far more than I ever expected to discover. It is a view that can live forever.

The Voice in the Willow Quarter

You put your car to sleep high on the hill. You have sauntered for nearly a mile through the uplands before descending to this small copse of willows by the river's edge. You have been asking the trees and the birds to tell you how they came to be in this valley and why they bother to keep going.

Your questions are old withering ones that expend far too much effort. The drive of blood and the dancing voice you have been after—the things that you suspect to be still alive in this rivervalley—are alive and well inside the stomping foot of your city. The valley is not really that different. Perhaps that is what you came here to discover.

Yes, you are listening to the voice of a Japanese wild fox spirit, a *yakozei,* or a Nahuatl *coyotl* trickster perhaps, or perhaps you are listening to the voice of the last bandit landscape. When you are ready to hear what I am really saying, perhaps it will be your own voice.

My voice is as fluid and as useless and as precious as water. It is the voice of one who has no explanation for anything. The only thing that I have found and kept is a way not tied to anything. I stuff my house in my shoes. And no matter how far this shanty house wanders, the journey is always homeward. I can be found in a thousand places. I am sitting on the head of Navaho Mountain, and I am walking the belly of the Henry Range. I am sitting at the foot of a glacial remnant in the far north. In one place in this icy foot I have found a mouth that is slowly burning open. I am listening there for a word. The only response I have received so far was a tooth of ice, as large as a granite house, exploding outward.

Yes, like you, I believe that there is an answer. But this answer cannot fit the shape of a word. When I have tried to cut up the earth or shape it whole into a word, the best that I have done is to cause a rain of ideas. My best perception is only a veil with dim figures in the distance. Perhaps they are nothing really, only ghosts of an idea.

I have found nothing but beginning. The only answers that seem to count look like a burning star or like a river coming around the bend in this valley or like the subtle nuance of a season. These answers can look like a fallen antler in late winter, and in spring like a speckled egg in a downy knothole.

I will recount to you a few of the particles of one rain-filled spring day in this willow grove alongside the river. Perhaps then you can discover the way in which this place is inside your from-where-you-come. With each of your steps along the upland path, you must have erased a raindrop from the storm in your eyes. The dance of the landscape must be slightly clearer in your eyes. I believe that you are attempting to go under a different sort of rain.

A leaf dives, just now, and a hat hits the ground. A rain cloud is shattering in the grass. But this cloud pulls itself together into a rainstream, and it is running out from between two fields of grass. It is running down to the river. The star mosses hold on for their dear lives. The river swells into the grass. Cane grass stems break off and drown. Only an old bone of elm stands up against all this melt.

It is an ice-cold, windful day. And yet it is spring. Winter is falling apart. The valley appears to be turning into one very icy river. Duck wings whir above this stream. Carp gills sway as slow as grass below it. Around a turn, a heron stands knee-deep in a cold that is just beginning to perk and simmer. All these lives are beginning to think about their own kind.

Above the river's dying ice, the last snowcrust drips and collapses. The changing forms of water sound from every nook and crevice. Each waking plant seems to have heard this bell. The ferns are itching to get up, but they hold onto their uncoiling heads for a few more weeks. Above this lowland the fox already has its milk-young. And higher still, where the river is the sky, a crow couple sing to one another, this bottle-shaped young man and young woman who come together to sleep. Their song as I first hear it is beyond words. It is a sharp cry, a force of goodwill, that comes banging against my ear. At one moment the crow fills this valley with a sharp sound, like stone cracking against stone. Just as suddenly the air is still. The river goes tawny and reflective. The rain sits down in a pond.

This ancient chamber, this earth house, is not yet half-made. It is a firm stage composed of the living and the inanimate, bonded to kin now melted to soil, and still deeper to ancestors, melted to stone. And yet this valley is one fluid stream of refreshing crow and water music, and it is just beginning one more annual eleven-billion-mile journey toward the brim of the cosmos. Nothing is finished. This place in this time is still near the dawn of creation.

In this time of spring migration there is always something enchanting about to begin. There is a sense of music in water. Trees swim in the wind. The air can be filled with words. There are words leaping inside the river and scurrying beneath smoldering humus. Their best life will never be in print. They cannot breathe inside that drying ink. It is the remarkable sense of presence in this valley that I especially appreciate. Everything is alive. This last breaking cloud that shattered in the grass—it smelled like fish. It had the rustle of silk. It had a wet rubber head. Its belly was on fire with lightning. It fell first on the back of a line of trees before it crashed down into the grass. It cut the sun in half. Yes, it was a rainstorm, but it was also a refined dance of real energy.

The rain from this last breaking cloud hit the forest like a hand passing through hair. It almost died in these wooden towers. But nothing really ends. This rain can do whatever it needs to do to live. In the blink of an eye it put on a froghouse mask, a pond. It hid in the ribs of the hill.

A bird, it is itself a rain of order that is far beyond me. I go under its sort of rain. It is a way that honors minor events. In its flurry no one event is of more importance and no one event is unimportant. Nothing goes unneeded.

I am without a comprehensive view. I have almost given away my eyes. My best response to any event is perhaps to try to show one small drop of awareness. At dawn, the night cold that has grown long and wild hoarfrost will suddenly cower in the grass and then be devoured by light. Dawn causes this change, and it gets here by itself. I have little to do with this dawn's work. I do not even stop to think about it. And when I presume myself to be beginning to understand, I only begin to doubt again.

Movement, plants, and stones are all there at the tip of my senses. In late winter the trees will bend with fresh snow. Rich opaque moonlight will stick like butter, spattered here and there. I can hardly wait to stream through this winter cluster of trees, and through summer's dust in the grass and its dance in the evening's backlight. Everything communicates.

My daily menu is simple enough. The wild sounds and textures come out to meet me. I do not have to go chasing after them. And I pay particular attention to the commonplace. In this cold rain I attend to the way in which the cascading drops sparkle like small fires. It is a hairsbreadth difference, and yet it is all the difference. I see this ghost-sun current that squirms out of these drops and moves snail-like in the arc of a rainbow.

Glance through the veil of these wiry willows. What is it that you behold? Yes, there is spring rain. And there is winter being burned alive. Seventy-three thousand tons of snow per square mile are running all over the place. Back and forth among the willows, the winter birds court. Collapse and drip—there are tea pools everywhere.

Spring is putting an end to one harsh winter one more time around. And yet harshness and terror can be found inside spring as well. There is usually more than one storm to recall. One such spring whirlwind that came up sent some bandaged clouds just ahead of it. They read like a sky full of dread. They were filled with terror. And when this storm came in, nothing, but nothing, could stop it. It put on a thick blue head. It lost its feet and bellowed like a madman. It was one mountain of air. It slapped the side of a coal town. A mouse pulled some earth over its head. And this copse of willows— it drew upon all its skill in bending and yielding.

As they give way to the storm, the willows also turn the light of the sun—this fusion bomb—into a soft meal. And from that fire and water and earth, the valley contrives small birds that move like hardened wind and rain pools and cracks in granite, the shade around stones, the erotic flowers, fifty-year-old "time-stains"—lichens—and the pungent damp earth. There is a cricket here and some others there. They seem to speak with legs sounding like ice.

Your city is far beyond this thicket. It is a jungle of stone and glass. It listens largely to words first. It always comes up smelling of scenery, too readily discarded and too thin— restrained to just one function. There is not enough air in its sky. It seems to fit the body of the earth like a torn shirt. It tries to be a head without a body, to live on ideas alone. And yet it is a landscape worth keeping. It is this willow copse that has made me begin to open to the possibilities of the city. I am no longer deceived by the city's water stiffening to rust and oil, or its old boxcars and mountains of scrap iron. The city is a scouring compass that points directly into the future. It has the potential to slow from a wild young river to become a living wilderness.

I am coming to see the city and the willow copse as doors to one rich and immense spiraling story. Now, having peopled the earth and having become residents more than migrants, we begin to look to where we stand. And now what I encounter here, in all local places, is not a played-out terrain or a remnant of the past or an artificial architecture. I find that I am inside a shimmering river of nature. The iron in a discarded nail has come bursting out of a supernova. It is a miracle.

The door to my future is in my cold spring hand. And what is so astonishing is that the pathway is not arduous. Listen. The wind shrills over the highest hill. It is plunging out of a washed-blue sky. The sound of this wind can be yellow on yellow, if I were to put it into a color today. And this wind-song can be even finer, such as the sound of wind against water or sand. It is not that different and no less important than the beating of your heart. At times in this willow copse it appears that there are private solos that are this landscape's alone, and far from the city and from my comprehension. And yet these too stand up with other sounds and colors if I continue to remain present. And there are momentous times when all the notes seem to be in perfect distinction and yet blended, and not incongruous with the music of the waking city.

Sound: I want to stay with its richness a while longer. Here comes another stanza of that rainsong: Just now there is one drop on the river. This bead leaves a deep, soft hole in the matt. I hear one drop, a fraction of an inch in diameter, hitting the middle of the river. I see it melting into rings of water dispersing outward. In this one small event trillions of molecules are shuffled around in a well-choreographed ballet. And as this rainfall increases, the river's surface tightens like a drumhead from dents everywhere at once. The sound builds to a high-pressure *cheeee*. Some of this rain has fallen from heights surpassing the Himalayas. They have fallen off a mountaintop to hit this drumhead.

There is music everywhere and of every tone. Willow leaves hiss as they cut the windstream into strips. And even in the autumn to come, when they leave the symphony of each tree, these leaves will continue to sing, just in a different manner. It is often a dry sound as they rustle in the grass or shatter underfoot. In any season I can find the decadence of rich black upon black or the opulence of a treasury or the bareness of a monk's cell. This color and sound that is our experience of a landscape can be a feast or a lean, starkly beautiful desert. The music of color and sound is as rich as the richest fantasy we might construct, and it is the *real* work of the earth. It is a tale that can never be fully told or sung, and, as ancient as it is, this tale is just beginning. Whenever I begin to think that I am running out of time, I always try to remember that this moment in my life is still inside creation. Perhaps more than anything else, it is this presumption that has enchanted my life.

I make my dream path out of these intricate sounds and the color of the worn-down hills and open day and deep night sky. It is a dream that is alive, and it is simple and gentle. It is a symphony that moves in dignity. I make my living dream out of these grasses of the grand prairie that rustle at ten decibels. My living dream is a homecoming. I am one small current in this vast chamber of color and sound. In this homecoming I believe that I begin to honor my life and to describe who I am.

*   *   *

You came to this place in this time to sense how a life ticks that lives with the willows. Well, maybe it's been *your* mind all the while. You likely wouldn't accept it, but maybe this willowy voice is a trickster, or, even worse, your own deluded, self-deceiving fabrication. My voice, it might just be this over-old land inside yourself. Perhaps you never even left your cottage in the city. Just because you won't believe that there could be that much room inside yourself means very little. Your reality is a consensus. It is too acquired. It is not the truth. Your childhood: Wasn't it a bend on a creek that feeds this river just north of here? There you did sit in the blistering sun. And in winter, once you bundled up and went walking about in a forest, in a killing blizzard, and you wept for joy. Why have you glossed over these monumental memories? Who is this old fox to whom you have been listening? Perhaps it is that broad, undiscovered terrain of *you* that is speaking. These words, they may just be your step-sounds when you are mowing the lawn or walking the dog.

You want to know where to look later for your master of the willow quarter? I will still be inside this copse, thrown away and entering like the rain of this day. And when you are back at your cottage in the city, look up. I will also be there—a brittle cloud, stuck to nothing. These events that appear to be only on the outside of you, because your eyes seem to be telling you this, just may be the landscape of you. And correspondingly, when you look into the mirror, it is the earth that may be dancing back in that smile of yours. Nothing is really that different. Each "thing" is really nothing, because it is no one thing, and more than its appearances, and ultimately perhaps everything. Close those two round deceivers for a moment, and the inside and the outside melt together.

You feel that your life is now far from the life of this willow quarter. You hear the tires rolling up and down the hill outside your bedroom, and the hiss of the freeway that is a quarter of a mile away. Every day, when you take your eyes for a walk, you believe that they are seeing a different story from mine. But for the immediate appearances, I am saying that perhaps everyone's story is the same one. The differences are insignificant.

You must realize now that you can find this place within your place. The baby in the cart, it will be this copse's brown speckled egg. And just look at your children becoming you beyond you. And even your cottage—the swish of straw on brick, a gush of wind, a tar roof and a garden stone, a long tail flickering in the elm—it is a sensate stream of energy. It must be so. How else could fiber and wood do such things? Your cottage, it is something magnificent at times, like this fogged-in willow copse. When you got up this day to set out on the drive to this river, one wall was colored by white blades of light. They were stronger than a bold thought. And inside your other small house, there was the thumping of the cosmos. There in your drab chest, there was nothing but sealed-up atomic fusion energy. You have always known it, but you have rarely permitted yourself to take the time to be simply present on the earth. Smack in the middle of the everyday, where you least expect to find it, there is *more* than magic. It is the work of the earth, looking in one place like a bird and in another place like a rolling tag of paper.

Return to this copse in another time. Your shirt will flutter in the wind. Your breath will feather and your hand will extend and form a teacup. A blade-thin flower of steam will rise from its hollow in the cool of the morning. It will be smoke-warm. Draw back this hand and taste. Perhaps it will be like the descent of cleansing rain. From this teacup you will read more than the story of a city. Your hand is of this river. It takes the form of a small wing, and it clenches at times like a thicket heaving and strained. It is one more small surprise that can turn your view topsy-turvy.

Come back to this grove when you are particularly thirsty.

There is no recession in this valley. Fish bellies bulge like the earth itself. A grasshead is swollen with seed, and there are a million others looking the same way. There is honey in wind. And your eyes shall feast on one solid character after another. Your ears will listen to short stories that are swelling into novels.

Come back to these willows when you find yourself in the middle of trying to do the impossible, when you are trying to be a river that is standing still. When you return, coat your ears with the clutter of leaves. Open your hand. Lean against a tree. You are beginning to trust that you can be anything when your thinking is untied. Your lips part like gills and you drink in a particle of wind before it runs under a cloud, and you awaken.